Ilham Zahir
Abdessamad BERAOUZ Mohamed B.
Ranya BOUKHRIS

Mode of action of bacterial toxins and toxin therapy

Ilham Zahir
Abdessamad BERAOUZ Mohamed B.
Ranya BOUKHRIS

Mode of action of bacterial toxins and toxin therapy

ScienciaScripts

Imprint

Any brand names and product names mentioned in this book are subject to trademark, brand or patent protection and are trademarks or registered trademarks of their respective holders. The use of brand names, product names, common names, trade names, product descriptions etc. even without a particular marking in this work is in no way to be construed to mean that such names may be regarded as unrestricted in respect of trademark and brand protection legislation and could thus be used by anyone.

Cover image: www.ingimage.com

This book is a translation from the original published under ISBN 978-620-6-72247-2.

Publisher:
Sciencia Scripts
is a trademark of
Dodo Books Indian Ocean Ltd. and OmniScriptum S.R.L publishing group

120 High Road, East Finchley, London, N2 9ED, United Kingdom
Str. Armeneasca 28/1, office 1, Chisinau MD-2012, Republic of Moldova, Europe
Printed at: see last page
ISBN: 978-620-8-26498-7

Foreword

❖ **Surname(s) and first name(s) of author(s) :**

- **ZAHIR** Ilham
- **BERAOUZ** Abdessamad
- **BERAOUZ** Mohamed
- **BOUKHRIS** Ranya

❖ **Job title :**

HOW BACTERIAL TOXINS WORK AND TOXIN THERAPY

❖ **Establishment :**

Polydisciplinary Faculty, Sultan Moulay Slimane University, Béni Mellal.

Morocco

❖ **Laboratory where the work was carried out :**

Laboratory of the Equipe Polyvalente en Recherche et Développement,
Department of Biology & Geology, Faculté Polydisciplinaire de Béni Mellal.

We dedicate this work :

To our parents, who today see their efforts and sacrifices rewarded. Through this report, they have watched over our education with infinite love and affection. We pray to Allah to grant them good health and long life. To our brothers and sisters, in testimony to the strong fraternal ties that unite us. To all our friends, without exception, and our colleagues in the Life Sciences department.

Summary

Toxins are powerful pathogenic factors, produced by certain bacteria, fungi, animals and plants, and involved in the drastic interactions of these pathogens with the host organism. Bacterial toxins were the first molecules identified as responsible for serious bacterial diseases in humans and animals.

Classically, bacterial toxins are divided into endotoxins and exotoxins. In fact, endotoxins are membrane compounds of Gram-negative bacteria, while exotoxins are secreted proteins that act both locally and at a distance from the site of bacterial colonization.

Bacterial toxins are extremely diverse in terms of size, structure, receptors, enzymatic activity and specific modes of action.

Recently, a major effort has been made to unravel the mechanisms of action of these molecules, which are highly active and responsible for inducing such severe symptoms in higher organisms. We speak of extracellular activity forming membrane pores and intracellular activity where many functional proteins are inhibited, such as G proteins, RAS proteins, translation elongation factor 2 and SNAREs proteins.

As a result, bacterial toxins are at the frontier of various disciplines, including bacteriology, cell biology, molecular biology, immunology and vaccinology. In this way, they are applied in a number of therapeutic and aesthetic fields, representing their positive side. This is what we call toxin therapy.

Key words: Bacteria; bacterial toxins; mechanism of action; toxin therapy.

ISTE OF ABBREVIATIONS

A. hydrophila:	*Aeromonas hydrophila*
AC :	Adenylate cyclase
ADP :	Adenosine diphosphate
cAMP :	Cyclic adenosine monophosphate
mRNA :	Messenger ribonucleic acid
rRNA :	Ribosomal ribonucleic acid
tRNA :	Ribonucleic transfer acid
ARTT :	ADP-ribosylating in turn
AT :	Tetanus toxoid
ATP :	Adenosine triphosphate
***B. pertussis* :**	*Bortedella pertussis*
BoNT :	Botulinum neurotoxin
C3bot :	*Clostridium botulinum* C3 exotoxin
***C. botulinum* :**	*Clostridium botulinum*
***C. diphteriae* :**	*Corynebactrium diphteriae*
C. tetani:	*Clostridium tetani*
Ca :$^{2+}$	Calcium ion
CDC :	Cholesterol-Dependent Cytolysins
Cl :$^{-}$	Chloride ion
MHC :	Major Histocompatibility Complex
CNF :	Cytotoxic Necrotizing Factors.
Factors)	
CT :	Cholera toxin (Toxine cholérique)
DT :	Diphtheria toxin
***E. coli* :**	*Escherichia coli*
EHEC :	Enterohemorrhagic *E. coli*
EHEC-hly :	Hemolysin from enterohemorrhagic *E. coli* strains
FE2 :	Elongation factor 2
GABA :	γ-aminobutyric acid (Gamma-aminobutyric acid)
GAP :	GTPase Activating Protein
Gb3 :	Globotriosyl ceramide
GDI :	GDP Dissociation Inhibitor. Inhibitor)

GDP :	Guanosine diphosphate
GEF :	GTPase Exchange Factors
GPI :	Glycosylphosphatidylinositol
GTP :	Guanosine triphosphate
Hlyα :	Hemolysin α
Hlyβ :	Hemolysin β
IgG :	Immunoglobulin type G
kDa :	Kilo dalton
***L. monocytogenes* :**	*Listeria monocytogenes*
LLO :	Listeriolysin O
LPS :	Lipopolysaccharide
Na :$^+$	Sodium ion
NAD :$^+$	Nicotinamide adenine dinucleotide
PFT :	Pore-forming toxin
pH :	Hydrogen Potential
PKA :	Protein kinase A
PTX :	Whooping cough toxin (Pertussis toxin)
PVL :	Panton& Valentine Leukocidin
RAS :	Rat sarcoma
GPCR :	G protein-coupled receptors
RTX:	Repeats in ToXins
***S. aureus* :**	*Staphylococcus aureus*
SAg :	Superantigen
TSS :	Toxic shock syndrome toxin
SDS-PAGE :	Polyacrylamide gel electrophoresis in the presence of SDS (Sodium dodecyl sulfate- polyacrylamide gel electrophoresis)
HUS :	Hemolytic Uremic Syndrome
AIDS :	Acquired Immunodeficiency Syndrome
SNARE :	Soluble N-ethylmaleinamid-sensitive factor Attachement protein Receptor
CNS :	Central Nervous System
PNS :	Peripheral Nervous System
Stx:	Shiga toxin
TCR :	T cell receptor

TeNT :   Tetanus neurotoxin

TSST :   Toxic Shock Syndrome Toxin   Toxic

shock syndrome toxin)

CONTENTS

INTRODUCTION

Bacteria are unicellular prokaryotic micro-organisms. According to the human-bacteria relationship, three types of bacteria can be distinguished:

Commensal bacteria are found either on the skin or in the mucous membranes of the body, feeding on organic matter without disturbing the organism. They play a role in stimulating the immune system and helping to activate vitamin B12 synthesis. However, in immunocompromised individuals (the elderly, infants, pregnant women, people with AIDS, diabetics, people on corticosteroids, etc.), or when the body's conditions change as a result of external factors such as lack of hygiene, cold trauma, injury or burns, commensal bacteria can become **opportunistic bacteria**, leading to infection. The third type of **microorganisms** are the **so-called "strict pathogens"**, whose presence in the body always reveals a pathology, whatever the state of the person's immunity: immunocompromised or immunocompetent.

Bacterial infection takes place in several stages. Unquestionably, once bacteria have crossed the body's natural barriers (skin and mucous membranes), they adhere to a tissue. They then multiply and grow, even invading certain cells.

At the same time, this infection triggers the body's immune system. Initially, a natural immune response is triggered by polynuclear cells, monocytes/macrophages and the complement system. In the second stage, a cell-mediated and/or humoral adaptive immune response is established. This leads to the production of specific antibodies with opsonizing properties.

However, some bacteria evade immune responses thanks to their high antigenic variability due to variation in structural genes, coding for a new variety of antigens not recognized by older antibodies. Bacteria can also hijack the immune system in order to ensure their survival by capsule or super-antigen production.

Thus, after **colonization, multiplication and escape from the immune system**, the ultimate stage in the infection of certain bacteria is the **production of toxic substances**, or toxins, which induce cytotoxicity in host cells. These virulence factors differ according to the target cells and their mode of action (membrane pore-forming toxins, ADP-ribosylating, proteolytic, etc.).

On the other hand, despite their harmful effects as powerful virulent factors, these toxins have **advantages for mankind**. We're talking about toxino-therapy, which aims to use bacterial toxins to prevent certain infections (anatoxins), just as they can be exploited as anti-cancer agents as well as in the aesthetic field.

The aim of this part of the course is to understand the effect of bacterial toxins on eukaryotic cells and the different modes of action by which these toxins act, as well as to study the advances made in the world of toxin therapy.

CHAPTER I: BACTERIAL TOXINS

A. Definition

Bacterial toxins are toxic, antigenic **substances** produced by certain **bacteria. At very low doses,** they can cause the **death of** a living organism (human, animal, plant, etc.) or induce **pathological disorders** in organs, tissues and cells.

They are either **protein-based**, encoded by genes, plasmids or phages on the bacterial chromosome, or **non-protein-based**: lipopolysaccharides (LPS), a constituent of the cell wall of Gram-negative bacteria.

B. Classification of bacteriotoxins

Bacterio-toxins can be classified in several ways: classically, clinically and according to their mode of action.

I. Classical classification

I.1 Endotoxins

Endotoxins (from the Greek: endon = inside and toxicon = poison) are LPS (figure 1). They are toxic biological substances released during bacterial lysis and, to a lesser extent, during multiplication. In the event of infection, they can cause a wide range of symptoms, from fever and chills to toxic shock.

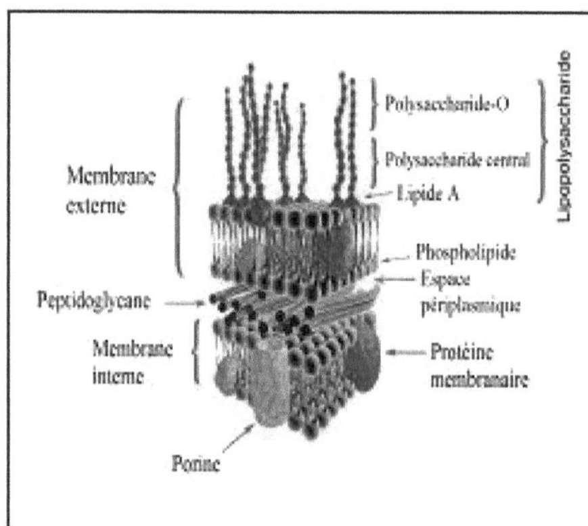

Figure 1 : Position des endotoxines au niveau de la paroi bactérienne à Gram négatif

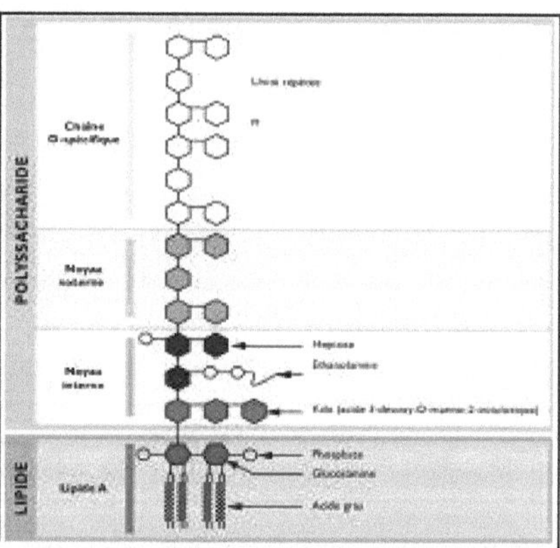

Figure 2 : Structure de l'endotoxine chez *Salmonella typhimurium*

Lipid A is the most conserved part of endotoxin. It is responsible for most of endotoxin's biological activity, i.e. its toxicity (Figure 2).

I.2. Exotoxins

Exotoxins are **soluble proteins** released **outside the** cell. Both Gram-positive and Gram-negative bacteria produce them.

Furthermore, they are toxins that spread from the site of infection to other parts of the body, causing **extreme damage** to the host by destroying cells or interfering with normal cellular metabolism.

I.3 Characterization of endotoxins and exotoxins

Several criteria can be used to distinguish between exotoxins and endotoxins (Table I).

Table I: Characteristics of endotoxins and exotoxins

Properties	Exotoxins	Endotoxins
Cell-toxin relationship	Released mainly outside the cell	Part of the bacterial cell
Origin	Gram-positive and Gram-negative bacteria	Gram-negative bacteria
Heat action	Thermolabile (except staphylococcal enterotoxin)	Thermostable
Toxic power	Very high	Moderate
Antigenic potency	Very high	Low
Vaccinating power	Very high	Low
Transformation into toxoids	Yes	No

II. Clinical classification of toxins

Clinical or epidemiological classification is a scientific method whose role is to classify toxins according to their action on a variety of target cells and the pathologies associated with these molecules.

II.1 Necrotic toxins

Necrotic toxins are toxins produced by several types of bacteria. They are capable of causing serious infections such as **necrosis**. Necrosis is a "disorderly" cell death which generally affects groups of cells. Take, for example, the case of dermo-necrotic toxin (DNT) from *Bordetella pertussis*.

II.2 Neurotoxic toxins

Neurotoxins are a group of toxins whose **highly specific effects on the nervous systems** of animals, including humans, interfere with the transmission of nerve impulses. They vary in chemical structure and mechanism of action, and produce very distinct biological effects.

The main toxins acting on the nervous system are the **clostridial toxins**: tetanus neurotoxin

(TeNT) and botulinum neurotoxin (BoNT) are proteins secreted by anaerobic bacteria of the *Clostridium* genus: *C. tetani* and *C. botulinum*. They are **lethal at very low doses**, making them the most powerful toxins known.

II.3 Hemolytic toxins

Bacterial hemolysins are substances capable of **destroying red blood cells**. They are certainly the easiest toxins to detect, since all it takes is a culture on blood agar to observe their effects. There are several hemolysins produced by many types of bacteria, including those of *E. coli*; shiga toxin, hemolysins α and β (Hly α and Hly β) and EHEC-hly (hemolysin of enterohemorrhagic strains of *E. coli*), hemolysin gamma of *Staphylococcus aureus* and CyaA toxin of *Bortedella pertussis*.

Hemolytic uremic syndrome (HUS) is caused by infection with shiga-toxin produced by Shigella and enterohemorrhagic strains of *E. coli*. It is a rare disease with hemolytic anemia, trombocytopenia and renal failure.

II.4. Leucoccidin toxins

Leukocidins are toxic substances produced by certain strains *of Staphylococcus* and *Streptococcus* that **destroy cells of the leukocyte lineage**. *Staphylococcus aureus* leukotoxins are a family of pore-forming toxins whose activating and lytic action is primarily directed against leukocytes.

In general, every toxin released by a bacterium can cause infection, leading to the establishment of the **bacterial-toxin-disease** relationship. The following table illustrates some examples:

Table II: Examples of bacterial-toxin-disease relationships

Bacterial toxin	Producing bacteria	Disease
Cholera toxin	*Vibrio cholerae*	Cholera
Botulinum toxin	*Clostridium. botulinum*	Botulism
Tetanus toxin	*Clostridium tetani*	Tetanus
Shiga toxin or verotoxin	*Shigella* and enterohaemorrhagic *E. coli* (EHEC; serotype: O157:H7)	Hemolytic uremic syndrome
Panton-Valentine Leukocidin (PVL)	*Staphylococcus aureus*	Pneumonia and nosocomial pneumonia
Diphtheria toxin (DT)	*Corynebacterium diphtheriae*	Diphtheria

RTX toxin: hemolysin A	*Escherichia coli*	Cystitis and pyelonephritis in the lower and upper urinary tract, respectively

III. Classification according to exotoxin mode of action

Bacterial toxins are best classified according to their mode of action. Some toxins act outside the cell, creating pores in the membrane. Other toxins are capable of inducing translocation of a catalytic fragment into the cytoplasm.

III.1 Action on the cell membrane

This group includes toxins with activity on the cytoplasmic membrane of eukaryotic cells: toxins of Gram-negative bacteria, also known as RTX *(Repeats in ToXins)* cytotoxins, and cholesterol-dependent cytolysins (CDC).

III.1.1. Pore-forming toxins of the RTX family

RTX toxins are produced by a range of Gram-negative bacteria. Their molecular mass fluctuates from 30 kDa to over 600 kDa, with cytotoxins being the best-known group, mainly due to the hemolytic activity of hemolysins. A basic feature of this large family of toxins is the presence of glycine- and aspartate-rich repeats. RTX toxins form aqueous pores of varying diameters.

III.1.1.1. Example of aerolysine

a. Definition

Aerolysin is a pore-forming toxin (PFT) secreted by *Aeromonas hydrophila.* It is the main cause of pathogenicity of this Gram-negative bacterium in amphibians, fish, reptiles and mammals, including man. This toxin is synthesized in the bacterium's cytoplasm in an inactive form known as prepro-aerolysin.

b. Chemical structure

Aerolysin is a protein made up of 470 amino acids. It is highly hydrophilic, consists essentially of β-sheets and is organized into at least three domains.

c. Mechanism of action

Prepro-aerolysin has a classical signal sequence which is cleaved as it passes through the bacterial cytoplasmic membrane. Pro-aerolysin is found in the periplasm and then crosses the wall via an export mechanism that is still poorly understood. Once in the extracellular environment, several steps are involved in the formation of the pore by aerolysin (Figure 3). (1) The water-soluble aerolysin precursor, forming dimers, can diffuse to the target cell, where it recognizes with high affinity the glycosylphosphatidylinositol (GPI) motif, which acts as the receptor on the outer side of the cytoplasmic membrane. (2) Activation by proteolytic cleavage of a C-terminal fragment of pro-aerolysin to aerolysin is achieved by a cellular protease, furin. (3) Activated aerolysin then becomes susceptible to oligomerization (assembly). (4) Insertion of the pore's oligomeric channels into the lipid bilayer follows oligomerization (assembly in the form of heptamerization in the case of aerolysin) and consequently breaks the permeability barrier of the cytoplasmic membrane, leading to osmotic lysis of the target cell.

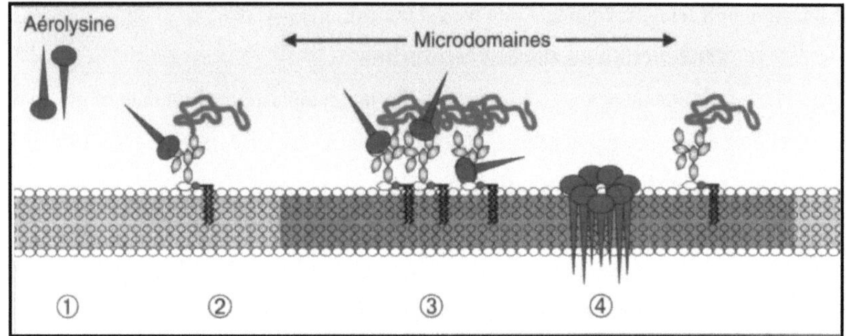

Figure 3: Pore-forming aerolysin assembly.

d. Diseases caused by *Aeromonas hydrophila*

The most common human infections caused by *A. hydrophila* are gastrointestinal, skin (cellulitis), wound and soft tissue infections, and bacteremia in immunocompromised individuals.

Symptoms of gastrointestinal infection range from watery diarrhea to dysenteric or bloody diarrhea. Chronic infection is also possible, and persistent intermittent diarrhea can progress to severe infection, sometimes months after the initial infection.

The bacterium in question has also been implicated as a cause of peritonitis, cholangitis, meningitis, septic arthritis, osteomyelitis, myositis, eye infections, urinary tract infections, nephropathy, septicemia, pneumonia and acute cholecystitis.

A. hydrophila is also the causal agent of certain aquatic animal diseases, notably the potentially fatal frog red leg disease, whose symptoms are attributable to the endotoxins and hemolysins produced by the bacterium.

III.1.2. Toxins of the cholesterol-dependent family

Cholesterol-interacting toxins, also known as cholesterol-dependent cytolysins

(CDCs), form a large family with over 20 members. These virulence factors are 50-60 kDa molecules produced by many Gram-positive bacteria, in particular perfringolysin from *Clostridium perfringens*, streptolysin O from *Streptococcus pyogenes*, pneumolysin from *S. pneumoniae* and listerolysin from *Listeria monocytogenes*. They feature a carboxy-terminal consensus motif of 11 amino acids, including a conserved cysteine. They oligomerize in the membrane to form very high molecular weight complexes. As their name suggests, CDCs require the presence of cholesterol to act. The choice of cholesterol among other host cell membrane lipids is important, as it ensures that the toxin cannot be harmful to the bacteria that produce it, since bacterial membranes lack cholesterol. CDCs are secreted by pathogenic bacteria as monomers in water-soluble form into the extracellular medium. These eventually bind to a receptor in the cholesterol-containing membrane of the target cell, where they oligomerize in a cycle of up to 50 monomers to form very large pores 300 Å in diameter. Thus, membrane insertion of CDCs leads to changes in ion flux across damaged membranes, culminating in cell lysis.

III.1.2.1. Example of listeriolysin

a. Definition

Listeriolysin O (LLO) is a soluble protein with a molecular weight of 56 kDa. It is secreted by *Listeria monocytogenes,* a Gram-positive pathogenic bacterium implicated in a food-borne infection known as listeriosis.

b. Crystal structure

LLO is an elongated rod-shaped molecule with four distinct domains, named D1 to D4 (Figure 4). D1 has a α / β fold with a five-stranded β leaflet and is surrounded by six α helices. D2 has four β helices and forms a three-strand antiparallel β leaflet connecting D1 to D4. D3 consists of a five-strand antiparallel β sheet, which is surrounded by six α helices in a α / β / α fold. D4 consists of two four-strand β sheets. While D1, D2 and D3 are intertwined, D4 appears as an independent holding unit connected to D2 via a glycine (G-417). The undecapeptide 483ECTGLAWEWWR493, which is highly conserved among CDCs, is located on the side of D4 opposite D2.

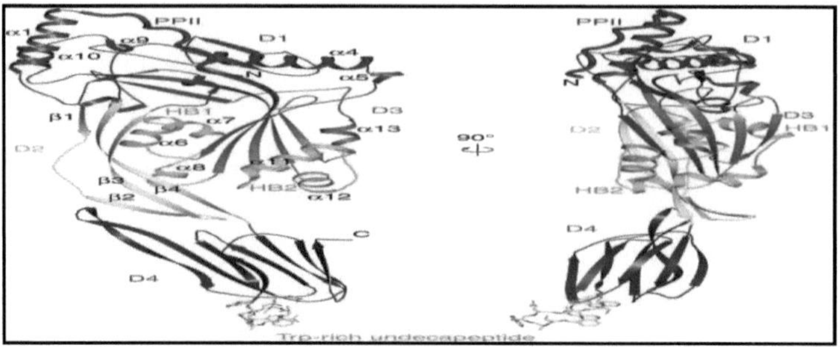

Figure 4: Crystal structure of listeriolysin O.

Individual domains are displayed in different colors. The PPII helix is shown in purple, D1 in red, D2 in yellow, D3 in green and D4 in blue. Membrane-inserted helix bundles (HB1 and HB2) in D3 are shown in cyan. The tryptophan-rich undecapeptide at the tip of D4 is shown as a rod model in green.

c. Mechanism of action

Studies show that LLO's efficacy and mode of action as a membrane-breaking agent are highly dependent on membrane, cholesterol content and environmental pH. LLO is capable of pore formation and membrane damage.

The toxin binds to cholesterol via two consecutive amino acids, Thr-515 and Leu-516. The latter can recognize the OH group of cholesterol and the first ring of its sterane backbone. In other words, the "cholesterol sensor" residue pair can enable the C-terminal part of soluble monomeric LLO to detect cholesterol in the target cell's plasma membrane and promote the first toxin-membrane contact. The apolar undecapeptide is responsible for the initial insertion of LLO into the target cell membrane (Figure 5A).

Furthermore, the cholesterol bound to the toxin's C-terminal domain has an extended linear conformation and does not present any inclination in LLO monomers inserted into the host cell membrane, so the orientation of the toxin remains perpendicular to the plane of the membrane, which is necessary for the formation of a large oligomeric pore. This edifice requires the oligomerization of 36 LLO monomers, which come into close contact thanks to the Leu-461 and Trp-489 linkage of two juxtaposed monomers (figure 5B).

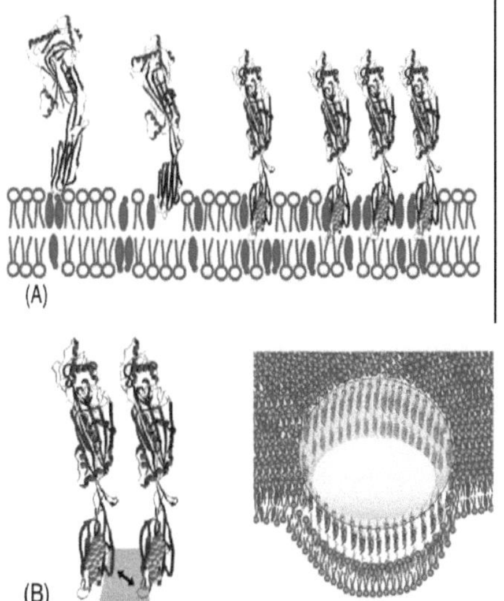

Figure 5: Insertion and formation of oligomeric pores by listeriolysin O.
(A): Insertion (from left to right) of LLO into the cholesterol-rich domain of the plasma membrane. Lipids are in green and cholesterol in red, but toxin-bound cholesterol molecules are shown in yellow. Initial contact of the soluble LLO monomers with the membrane is caused by the surface binding of the Thr-515/Leu-516 "sensor" to the accessible parts of the cholesterol. (B): Schematic drawing of a ring of 36 LLO monomers. The oligomerization process involves close contacts between Leu-461 and Trp-489 (arrow) belonging to two vicinal monomers.

d. Listeriosis

Listeria monocytogenes enters the body through the digestive tract, via the ingestion of contaminated food. From the intestine, the bacteria spread to the lymph nodes and then to the bloodstream. The bacteria then multiply in the liver and spleen, the target organs for the bacteria. In most cases, the immune system controls the infection in immunocompetent subjects with an inapparent infection. In such cases, gastroenteritis has been described, with diarrhea occurring a few hours after eating contaminated food, and usually without neurological complications or scpticcmia.

However, if the inoculum is massive, or in certain fragile subjects, the infection is not controlled in the intestine, spleen and liver.

In fact, the bacterium adheres to target cells via internalin, which interacts with E-cadherin, a receptor expressed on certain cells (intestinal...), thus inducing endocytosis. The microorganism **destroys the endocytosis vesicle** and moves into the cytoplasm under the action of listerolysin O and phospholipase C. In the cytoplasm, the bacterium polymerizes actin, creating comets that propel the bacterium out of the cell, enabling it to **spread** from one cell to another (figure 6).

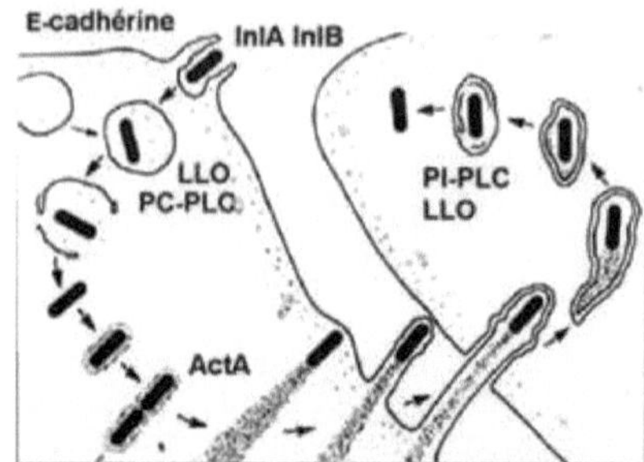

Figure 6: Creation of actin "comets" by *Listeria monocytogenes*
Act: actin; Inl: internalin; LLO: listeriolysin O; PLC: phospholipase C

Once released into the bloodstream, the bacterium causes **listeriosis**, a form of **septicemia** with a variety of clinical manifestations that occur mainly in fragile patients (pregnant women, immunocompromised patients undergoing chemotherapy, AIDS patients or those with liver abnormalities such as cirrhosis or hemochromatosis, or in certain genetically predisposed subjects).

In pregnant women, for example, *Listeria monocytogenes* targets the placenta and central nervous system, leading to **feto-placental infection**. In fact, after the fifth month of pregnancy, the bacterium colonizes the placenta, forming numerous inflammatory granulomas, leading to **chorio-amniotitis** and infection of the fetus *in utero* (90% of cases), causing meningoencephalitis with fever, headache, stiff neck and cranial nerve paralysis (rhombencephalitis). **Feto-placental infection** can often result **in premature delivery** after 5ème months, or in **spontaneous or repeated abortions** before that date.

Maternal-infantile infection is manifested by a flu-like syndrome with fever, chills, fatigue, headaches and myalgias that precedes delivery by 2 to 14 days or more.

Furthermore, if the fetus survives the infection *in utero*, the newborn will suffer cyanosis, apnea, respiratory distress, consciousness disorders (apathy, convulsions) and pneumonia from birth.

A febrile relapse with bacteremia is often observed in the mother during childbirth.

In less than 10% of cases, the newborn is infected in the perinatal period or during delivery, without placental infection. The child is born apparently healthy, and infection appears 8 to 60 days after delivery, with purulent meningitis, fever, insomnia, irritability and consciousness disorders.

III.2 Intracellular action

This group includes toxins with intracellular activity, also known as AB toxins. These toxins are generally organized into <u>two domains</u>: a catalytic A domain responsible for toxic activity, and a B domain which enables attachment to the target cell.

III.2.1. Toxins acting on the G protein

G (GTP-binding) proteins are a family of homologous proteins that couple to their cellular effectors. G proteins are composed of three subunits: the α subunit (39 to 52 kDa), the β subunit (35 and 36 kDa) and the γ subunit (around 5 to 8 kDa). G proteins can be divided into different groups, the most studied being stimulatory G proteins (G_S) and inhibitory G proteins (G)$_i$

G protein-acting toxins are toxins that interact with the G protein as intermediaries in the transmission of the effector part of the toxin from the outside to the inside of the cell.

In fact, these toxins are ADP-ribosyltransferase enzymes that catalyze the transfer of ADP-ribose from NAD^+ to the α-subunit of the G protein. Thus, these toxins can covalently modify the α-subunits of certain G proteins. Examples include the bacterial toxin of *Vibrio cholerae* (CT), which activates G_S , and that of *Bordetella pertussis* (PTX), which inhibits coupling between G proteins and their associated receptors.

III.2.1.1. Toxins acting on the Gs protein

III.2.1.1.1. Example of cholera toxin

a. Definition

Cholera toxin (CT), produced by the bacterium *Vibrio cholerae*, is an exotoxin, a toxic protein excreted into the external environment. These are excreted and assembled in the periplasm, then excreted into the extracellular environment.

b. Chemical structure

Cholera toxin is an 84 kDa protein made up of five B subunits (56 kDa, each with 103 amino acids) and one A subunit (240 amino acids, 28 kDa). The latter is made up of two fractions, $\Lambda1$ and $\Lambda2$, linked by a disulfide bridge.

The A and B subunits are encoded by the *ctxA* and *ctxB* genes, located in the same operon under the control of the same promoter.

c. Mechanism of action

Cholera toxin exerts its toxic activity by binding, via its B subunits, to a specific membrane receptor (ganglioside GMi) on the surface of enterocytes, but also on numerous other cell lines (Figure 7).

This binding induces the rapid translocation of the toxin's enzymatic A subunit across the membrane. The A1 subunit catalyzes the hydrolysis of NAD and the transfer of part of this molecule (ADP-ribose) to the regulatory subunit (Gα_S) located on the inner surface of the plasma membrane.

In summary, ADP-ribosylation of this regulatory protein leads to permanent activation of adenylatecyclase, and uncontrolled conversion of ATP to cAMP. The increase in intracellular cAMP levels leads to active chlorine secretion, and inhibits the joint reabsorption of this ion and sodium by the enterocyte. This mechanism, which is non-lethal to the cell, leads to a passive outflow of water, the cause of diarrhea.

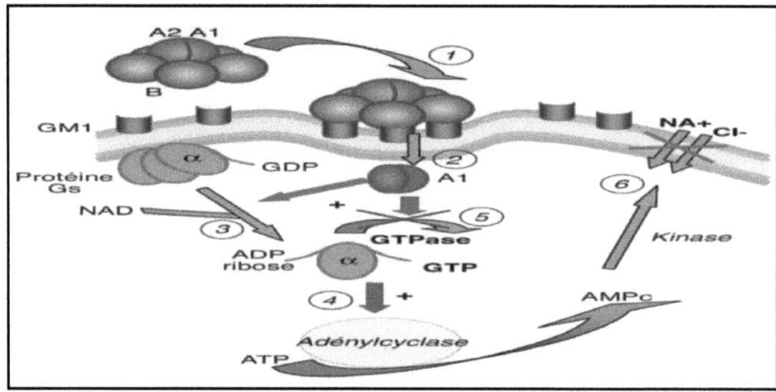

Figure 7: Cholera toxin activity.
The toxin released into the intestinal lumen binds to GM1 gangliosides on the surface of epithelial cells (1). Only the A1 subunit penetrates the cell interior (2). It catalyzes the transfer of adenosine diphosphate (ADP) ribose from nicotinamide adenine dinucleotide (NAD) to the α subunit of the Gs protein. Guanosine diphosphate (GDP) is converted to guanosine triphosphate (GTP) (3). The combination of ribosylated α-ADP and GTP is a potent activator of adenylcyclase (4). In addition, the A1 subunit inhibits the natural intrinsic GTPase activity of the Gs protein, preventing the return to the inactive Gs-GDP protein form (5). Activation of adenylcyclase leads to the production of large amounts of cyclic adenosine monophosphate (cAMP), responsible for kinase-mediated arrest of Na$^+$ and Cl$^-$ ion reabsorption (6).

d. Cholera

Cholera is a clinical syndrome that manifests itself in the form of epidemics affecting developing countries, and more specifically the poorest and most vulnerable populations.

Cholera is caused by ingestion of food or water contaminated with an infectious dose of *Vibrio cholerae*. After ingestion, the bacteria pass through the stomach into the small intestine. They then attach themselves to the cells of the intestinal mucosa, producing the cholera toxin.

The result is **cholera**, characterized by the appearance of abundant acute watery diarrhea in the form of "rice water" (figure 8).

If not treated promptly, this can lead to acute dehydration, circulatory collapse, renal failure and death.

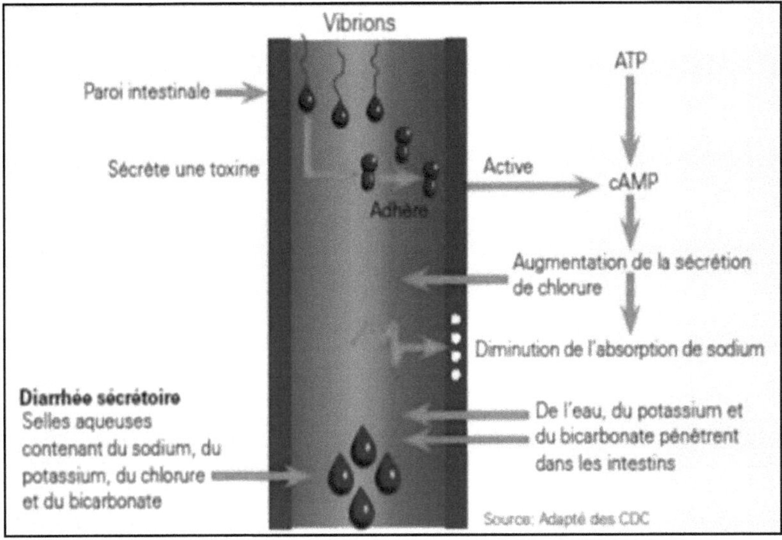

Figure 8: Pathophysiology of cholera.

III.2.1.2. Toxins acting on the Gi protein
III.2.1.2.1. Example of PTX
a. Definition

Pertussis toxin (PTX) is produced by *Bortedella pertussis* (*B. pertussis*) and belongs to the ADP-ribosylation family of bacterial exotoxins.

b. Crystal structure

The structure was determined in 1994. In fact, it is a complex ADP-ribosylating toxin composed of six different subunits (figure 9), and arranged in the A-B architecture. Protomer A consists of a single S1 subunit, while oligomer B comprises S2, S3, S5 and two S4 subunits. The A protomer catalyzes the ADP-ribosylation of a cysteine residue in the α subunit of the Gi heterotrimeric protein subfamily, while the B oligomer is responsible for binding to specific cell-surface receptors and releasing the A protomer into receptor cells.

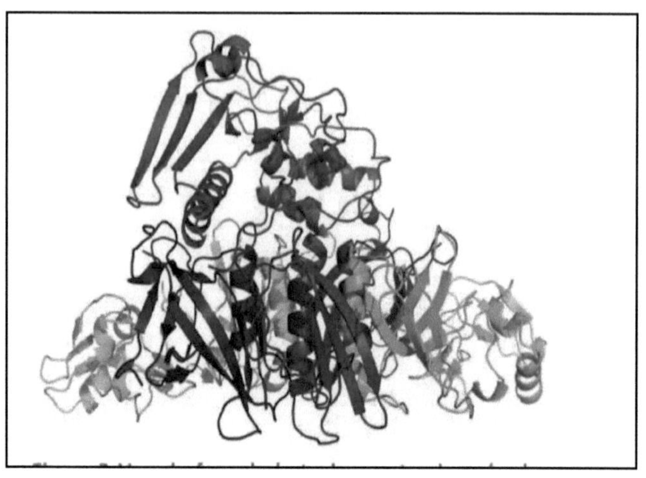

Figure 9: Front view of pertussis toxin.
In red the S1 sub-unit, in green S2, in cyan S3, in blue the two S4 sub-units
and in violet S5.

c. Mechanism of action

Following PTX binding to a host cell receptor via the S2 and S3 subunits of the B oligomer, the A protomer (S1 subunit) penetrates the membrane and detaches from the B oligomer in the cytoplasm. Once inside the cell, the A domain ribosylates specific target proteins such as the α-subunit of heterotrimeric Gi proteins through its ADP-ribosyltransferase activity. PTX catalyzes cleavage of the C-N bond between the ribose carbon atom and the nicotinamide nitrogen atom, transferring the ADP-ribosyl half of nicotinamide adenine dinucleotide (NAD^+) to an acceptor molecule on the target protein. In the case of PTX, ADP-ribosylation of Gα proteins$_i$ prevents coupling to their G-protein-coupled receptors (GPCRs) and thus disrupts the signal transduction cascade.

Uncoupling of the GPCR from the Gα proteins$_i$ leads to disruption of communication between the receptor and the effector molecule, **adenylatecyclase (AC)**. As a result, the Gα protein$_i$ is inactivated and can no longer perform its normal function of inhibiting AC. In this way, it prevents the GPCR signal$_i$. As a result, the conversion of ATP to cAMP cannot be interrupted, leading to excess intracellular cAMP levels and subsequent disruption of many cellular processes, as shown in figure 10.

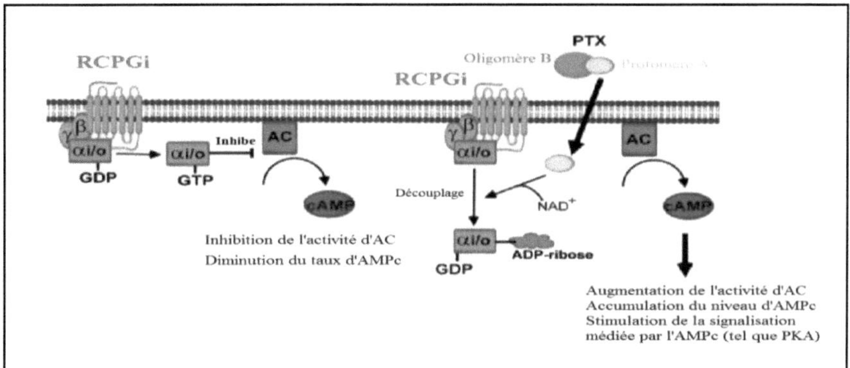

Figure 10: Uncoupling of the Gα protein$_i$ from their GPCR.

d. Whooping cough

Pertussis is a bacterial infection of the lower respiratory tract with little or no fever, but with a long and highly contagious course, caused by PTX produced by the bacterium *B. pertussis*.

The first phase is colonization. After inhalation, *B. pertussis* settles in the trachea or bronchi, attaching itself to the tracheal ciliated cells via adhesins (more precisely, to the vibratory cilia, causing their paralysis). During this phase, the severity and duration of the disease can be reduced by early antimicrobial treatment.

The second stage is the toxemic phase. Symptoms result from the microorganism's interaction with the host's immune system. The bacterium multiplies outside the tissues and synthesizes toxins that act together to destroy ciliated cells, causing an accumulation of mucus. These toxins also act on a systemic level, leading to an increase in lymphocytes. During this phase, *B. pertussis* infection becomes difficult to control with antimicrobial agents.

III.2.2. Toxins acting on Rho proteins

III.2.2.1 Overview of Rho proteins

Rho proteins belong to the small GTPase-Ras superfamily; among the best-known are RhoA.

There are two distinct Rho protein conformations: **active** GTP-binding and **inactive** GDP-binding. The change of configuration is regulated by **GEF** (GTPase Exchange Factors) proteins, which are exchange factors that promote GDP dissociation and GTP binding to the Rho protein, and by **GDI** (GDP Dissociation Inhibitor) and **GAP** (GTPase Activiting Protein) proteins, which promote the return of Rho proteins to their inactive state (Figure 11).

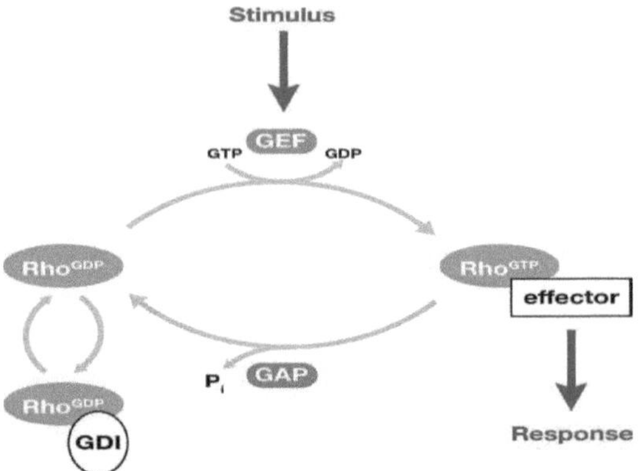

Figure 11: Rho protein cycle

In the basal state, the RhoA protein in GDP form is associated with the cytosolic factor Rho-GDI. RhoA is converted to its GTP form by GEF exchange factors. In its GTP form, the protein is able to interact with its effectors.

Small Rho cells return to their basal state by hydrolysis of GTP to GDP, due to their intrinsic GTPase activity, stimulated by GAP proteins.

The Rho protein is involved in controlling the organization of the actin cytoskeleton. In fact, activated Rho seems to have two main targets: a serine/threonine kinase known as Rho kinase, and a kinase that transforms phosphatidyl inositol-4-phosphate by phosphorylation (at position 5 on inositol), into PI-4,5-P or PIP2. Indeed, by modulating these two activities, Rho could reorganize the actin cytoskeleton as follows:

Action on Rho-kinase causes phosphorylation of the light chain of type 2 myosins, enabling these molecules to associate with actin fibers and trigger contraction. By raising the concentration of PIP2, it seems that Rho can induce actin polymerization. Rho's second activity is to expel so-called "*capping*" proteins (formin) associated with the "bearded" end of actin fibers, where new actin subunits are added (figure 12). These two activities enable Rho to extend the cell surface by polymerizing actin fibers at the cell periphery.

Figure 12: Addition of new actin molecules to an actin filament
Rho activates formin to promote linear elongation of filaments with barbed ends.
Profilin is an actin-binding protein involved in the dynamic renewal and reconstruction of the actin cytoskeleton. It catalyzes the ADP-to-ATP exchange required for actin polymerization. In fact, actin monomers need to be recharged with ATP before joining the growing filament.

In addition, Rho induces the formation of adhesion points, which are the sites where the cell anchors itself to the extracellular matrix via integrins. This anchoring by integrins to the extracellular matrix is an essential step in the generation of intracellular signals leading to either cell multiplication or differentiation.

Clostridium botulinun exoenzyme C3 and *Escherichia coli* cytotoxic necrotizing factor (CNFl) are two toxins with opposite effects on the Rho protein. While the former inhibits Rho, the latter permanently activates it.

III.2.2.2 Toxins inhibiting Rho binding to its effectors

III.2.2.2.1. Example of C3bot exotoxin

a. Definition

Clostridium botulinum C3 exotoxin (C3bot) is not a true toxin, but only the 25 kDa enzymatic fraction of a toxin. This fraction inhibits Rho activity by transferring a portion of ADP ribose from NAD to small Rho GTPases. It is thus an ADP-ribosyltransferase.

b. Chemical structure

C3bot exotoxin is a 25 kDa, 251 amino acid single-chain protein containing the ADP-ribosylation site. Catalytic function and substrate recognition are ensured by the ARTT turn motif. A conserved β-structure core is assembled by joining β5 to β6, which anchors the ARTT motif. The NAD-binding site in the C3bot structure is flanked by a β-sheet core and an α3 helix (figure 13).

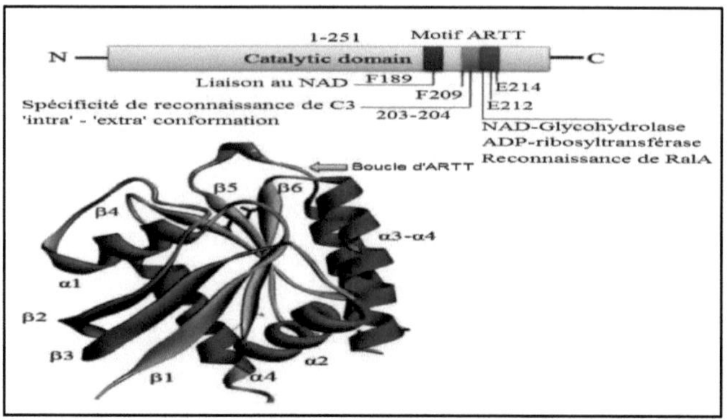

Figure 13: Structure and function of *C. botulinum* C3 toxin.

c. Mechanism of action

The *C. botulinum*   C3 exoenzyme is responsible for inactivating members of the Rho GTPase family involved in the reorganization of the actin cytoskeleton.

In fact, C3bot has a potential pathophysiological role towards immune cells, and more specifically monocytes and macrophages. It enters the cytosol of these cells via a specific endocytotic mechanism. The C3bot enzyme specifically mono-ADP-ribosylates Rho isoforms at asparagine-41 and prefers Rho-GDP as a substrate, because in this structure, Asn-41 is accessible to C3bot.

As a result, ADP-ribosylated Rho-GDP binds more efficiently to GDI, which traps Rho in Rho-GDI complexes in the cytosol. This prevents the translocation of cytosolic Rho to the cytoplasmic membrane and consequently inhibits its activation to Rho-GTP, as well as proscribing the subsequent activation of the various cellular Rho-effector molecules involved in many cellular phenomena.

For example, C3 blocks the cell's entry into the cell cycle and the disorganization of actin fibers (Figure 14), which is the most spectacular consequence of intoxication with the *Clostridium botulinum* C3 exoenzyme.

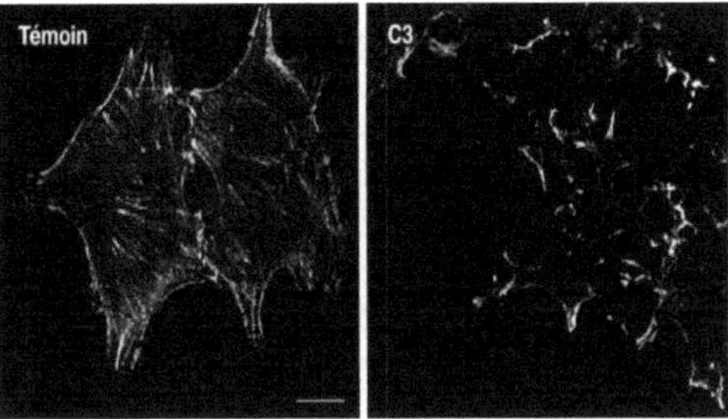

Figure 14: Effect of *Clostridium botulinum* C3 exoenzyme inhibiting Rho protein on the actin cytoskeleton.
Labeling with fluorescent phalloidin allows particularly clear observation of C3 exoenzyme-induced remodeling of the actin cytoskeleton (bar: 10 μm).

III.2.2.3 Toxins activating Rho binding to its effectors

III.2.2.3.1. Example of CNFl toxin

a. Definition

Cytotoxic necrotizing factor (CNF) is a toxin produced by certain strains of uropathogenic and enteropathogenic *Escherichia coli*. The toxin has a specific deamidase-type catalytic activity.

b. Chemical structure

Cytotoxic necrotizing factor (CNFl) is a 110 kDa protein. The toxin is composed of three domains: a catalytic domain, a cell-binding domain and a membrane translocation domain (figure 15).

Figure 15: Structure of CNF1 toxin
B: cell-binding domain; T: membrane translocation domain; C: catalytic domain causing deamidation of glutamine 63 from Rho GTP ases.

c. Mechanism of action

Once released by the bacterium, CNF1 toxin binds to a receptor on a uroepithelial host cell (1) and is then endocytosed (2 + 3). The toxin is transferred to a low pH endosomal compartment (4) where the catalytic domain is injected into the cytosol (translocation) (5). In fact, CNFl possesses a deamidase-type catalytic activity specific to Rho glutamine 63 (6). By deamidating glutamine 63 to glutamic acid, CNFl induces a mutation in Rho, permanently activating it (Figure 16). In effect, Rho loses its GTPase activity, both basally and when stimulated by GAP. In other words, under the effect of CNF1, Rho loses its ability to hydrolyze

GTP to GDP, either alone or assisted by GAP, which becomes blocked. Rho is thus said to be dominant-positive, with no need for interaction with its exchange factor.

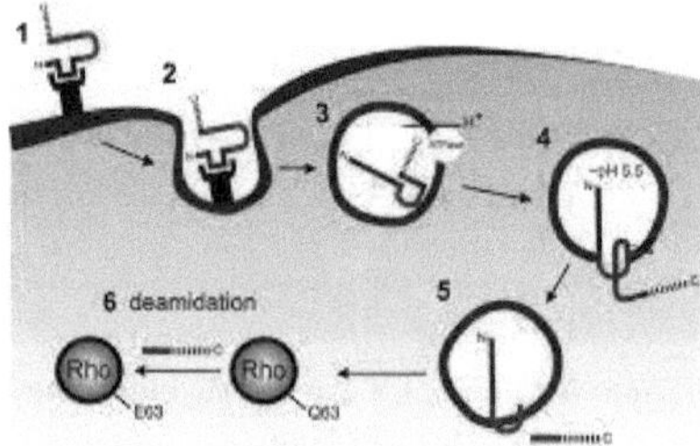

Figure 16: CNF1 mode of action

As a result, CNFl has diametrically opposed effects on cells to those of C3 (Table III). Unquestionably, in cells sensitive to it, it increases the number of tension fibers (actin contractile bundles) and the appearance of remodeling of the actin cytoskeleton. In this way, it induces a multiplication of adhesion focal points (Figure 17). In addition, it causes a marked increase in endocellular PI-4, 5-P kinase activity, as well as cell proliferation and phosphorylation of FAK (*focal adhesion kinase*) and paxillin proteins localized in adhesion sites.

Table III: Cellular activity of C3 and CNF1

Cellular activity	C3	CNF1
Actin tension fibers	Disorganization	Organization
Membership focal points	Disappearance	Training
p125 FAK	Dephosphorylation	Phosphorylation
Paxillin	Dephosphorylation	Phosphorylation
Myosin 2 light chain	Dephosphorylation	Phosphorylation
Smooth muscle contraction	Inhibition	Activation
Pinocytosis	Inhibition	Activation
DNA synthesis	Inhibition	Activation
Entry into S phase of the cell cycle	Blocking	Activation
Action on Rho	Inhibition	Activation
Enzymatic activity on Rho	ADP-ribosylation	Deamidation
Modified Rho residue	Asparagine 41	Glutamine 63

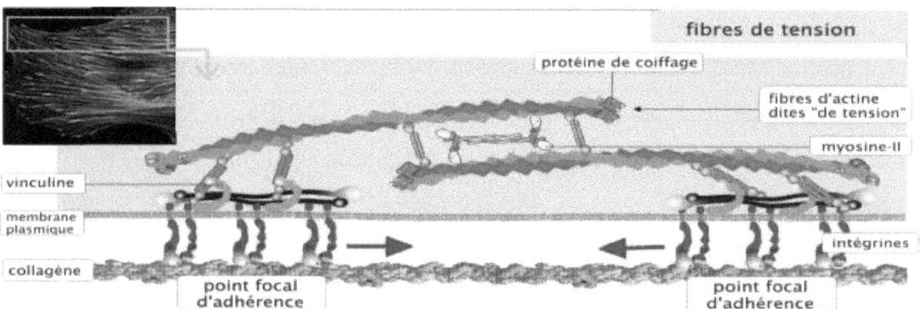

Figure 17: Tension fibers

III.2.3. Toxins acting on protein synthesis

The various stages of protein synthesis involve numerous molecular players, including messenger RNAs (mRNAs), ribosomes, transfer RNAs (tRNAs), synthetases and translation factors. These players are successively called upon during three major phases of translation: initiation, elongation and termination. Protein synthesis can be affected by certain bacterial toxins, such as diphtheria toxin, which blocks the activity of elongation factor 2, and Shiga-toxins, which act on 28S ribosomal RNA.

III.2.3.1. Action on elongation factor 2

Translational elongation factors are the workhorses of protein synthesis on the ribosome. Eukaryotic elongation factor 2 (eFE-2) is a 100 kDa protein that facilitates the movement of the peptidyl-RNAt-mRNA complex from the A site of the ribosome to the P site during protein synthesis. It is also responsible for the GTP-dependent ribosome translocation step. FE2 catalyzes the hydrolysis of GTP required for ribosome movement on mRNA, enabling protein synthesis.

In position 699 of its peptide sequence, FE-2 also contains a diphthamide residue, (2-[3-carboxyamido-3-(trimethylammnio)propyl]histidine).

III.2.3.1.1. Example of diphtheria toxin

a. Definition

Diphtheria toxin (DT) is one of the most extensively studied bacterial toxins, secreted by toxic strains of the bacterium *Corynebacterium diphtheriae* lysogenized with a phage carrying the DT gene. This toxin is responsible for the main symptoms and mortality of diphtheria.

b. Chemical structure

DT is a large molecule of 62 kDa. It is synthesized as a single polypeptide chain of 560 amino acids, encoded by the *tox* gene carried by a family of corynebacteriophages, under the control of a *P-tox* promoter.

A signal peptide of 25 amino acids is removed during secretion across the bacterial plasma membrane, resulting in a protein toxin of 535 amino acids. But in its active form, it is proteolysed to two polypeptide chains linked by a disulfide bond (figure 18). The C-terminal fragment "fragment B" (345 residues, 38 kDa) contains the transmembrane binding (T) and receptor (R) domains, thus facilitating the penetration of fragment A. Whereas the N-terminal fragment "fragment A" (190 residues, 24 kDa) is thermostable at 100°C, stable to pH variations and the action of proteolytic enzymes, and contains the catalytic domain (C).

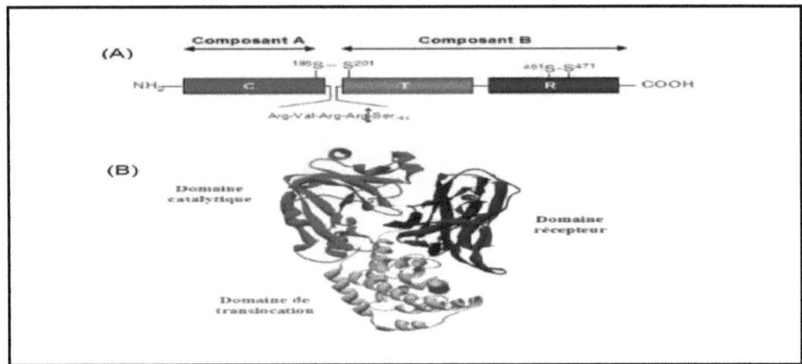

Figure 18: Structure of diphtheria toxin.
(A): Schematic representation of DT's three structural domains. The site of furin cleavage between the two fragments A and B is represented by an arrow. (B): Three-dimensional structure of DT.

c. Mechanism of action

DT is an ADP ribosyl transferase actively directed against the translational elongation factor 2 (FE2). Through an ADP-ribosylation reaction of FE2, DT inhibits protein synthesis (Figure 19). This process takes place in two main stages:

❖ *Translocation and activation*

When a cell is poisoned, DT binds to its cell-surface receptor via the R domain. The loop between Cyc-186 and Cyc-201 is then cleaved by furin, a cellular protease. DT is then endocytosed. Acidification of the endosome causes a major conformational change, enabling the T-domain α-helices to penetrate the endosomal membrane, forming a channel for translocation of fragment A to the cytosol. In its membrane-inserted form, the T domain partially interacts with the C domain, facilitating its unfolding during translocation. The C domain then passes into the cytosol via cytosolic factors, where it is restructured into its active form, then released by disulfide bridge reduction.

❖ *Catalytic activity*

The C -domain catalytic site of an ADP-ribosyl-transferase activity becomes free and consequently targets the NAD cofactor[+] . As a result, His-21 and Tyr-65 of the catalytic site are involved in NAD binding[+] , and Glu-148 plays a key role in catalysis, transferring a molecule of ribosylated ADP to a diphthamide residue of the translational EF-2. ADP-ribosylation of this residue leads to an arrest in protein synthesis, and consequent cell death (necrosis).

ADP-ribosylation of FE-2 is irreversible. No endogenous mechanism is capable of restoring FE-2 function. The absence of a mechanism to repair the damage caused by the C domain is probably at the root of DT's extreme toxicity.

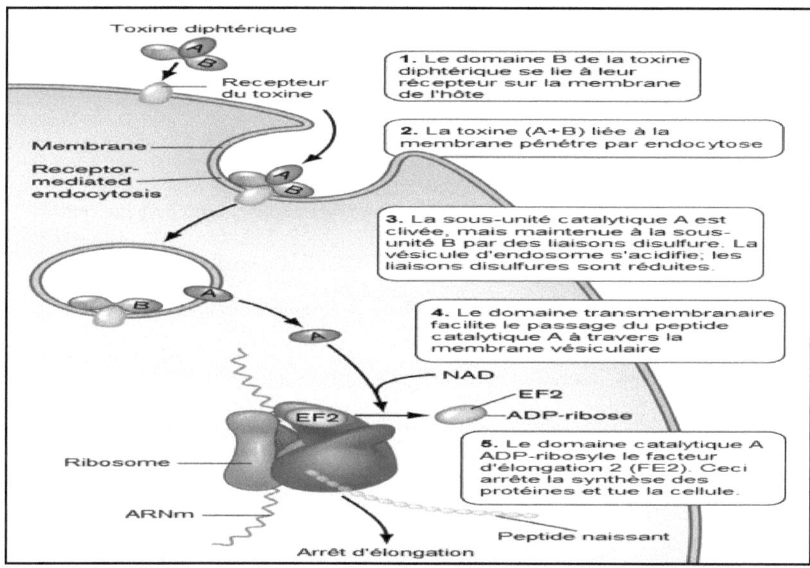

Figure 19: Schematic representation of the mode of action of diphtheria toxin.

d. Diphtheria

Diphtheria is an infection caused by the bacterium *C. diphtheriae*. Signs and symptoms usually appear 2 to 5 days after exposure, and range in severity from mild to severe. Symptoms often appear gradually, starting with a sore throat and fever. In severe cases, the bacterium produces diphtheria toxin, which most often causes a thick grey or white patch at the back of the throat: membranous pharyngitis (figure 20). This can block the airways, making it difficult to breathe or swallow, and can also cause a hacking cough. The neck may swell in part due to swollen lymph nodes.

The toxin can also enter the bloodstream, leading to complications such as inflammation, heart muscle damage, nerve inflammation, kidney problems and bleeding due to low platelet counts. Damage to the heart muscle can lead to an abnormal heart rate, while inflammation of the nerves can cause paralysis.

Transmission occurs via the airborne route through direct contact with patients.

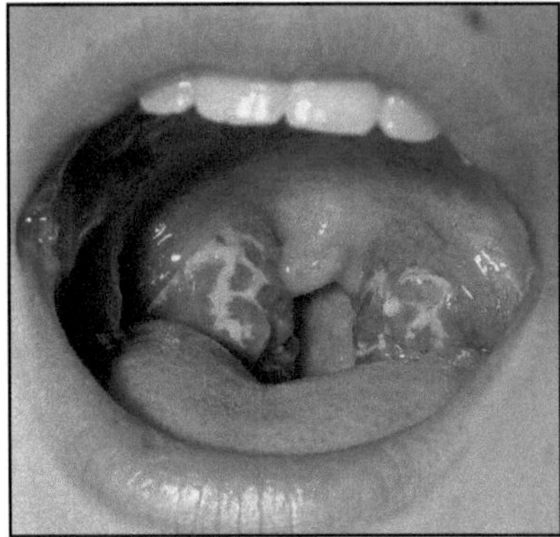

Figure 20: Pathogenesis of diphtheria: pharyngeal pseudomembrane.

III.2.3.2. Action at 28S rRNA level

rRNAs are the predominant components of ribosomes in eukaryotes. There are four different rRNAs: 5S, 18S, 5.8S and 28S, derived from a common 45S and 5S precursor.

The genes coding for these rRNAs are called rDNAs. 18S, 5.8S and 28S genes are in the same transcription unit of the 45S rDNA present as a tandem repeat in a genome, while 5S genes are organized in clusters of tandem repeats separated by small non-transcribed spacers.

Human 28S rRNA is 5025 base pairs long. Along with 5S rRNA, 5.8S rRNA and around 45 proteins, it is the main component of the large 60S subunit of eukaryotic ribosomes.

III.2.3.2.1 Example of Shiga-toxin 1

a. Definition

Shiga toxins (Stxs) or verotoxins are a group of bacterial AB5 proteins of around 70 kDa that inhibit protein synthesis in eukaryotic cells. Stxs block protein synthesis by removing an adenine residue from the 28S rRNA of the 60S ribosome. Stx1 is one of the most potent bacterial toxins known, produced by *Shigella dysenteriae* 1 and certain serotypes of *E. coli*. In addition to Stx1, certain strains *of E. coli* produce a second type of Stx: Stx2.

b. Chemical structure

Stx1 toxins are AB-type toxins consisting of an A subunit responsible for toxic activity (293 amino acids trypsin-sensitive in the "aa-248-251" region) which is asymmetrically cleaved to give two disulfide-bonded A1 and A2 subunits, and five identical B subunits (each with 69 amino acids) (Figure 21) enabling toxin binding to its specific glycolipid receptor; globotriaosylceramide (Gb3), an endothelial cell membrane component.

The genes for these subunits are located in the same operon, with the A subunit gene on the 5' side of the B subunit gene. This operon is generally found in the sequence of an inducible, lysogenic, lambda-type bacteriophage.

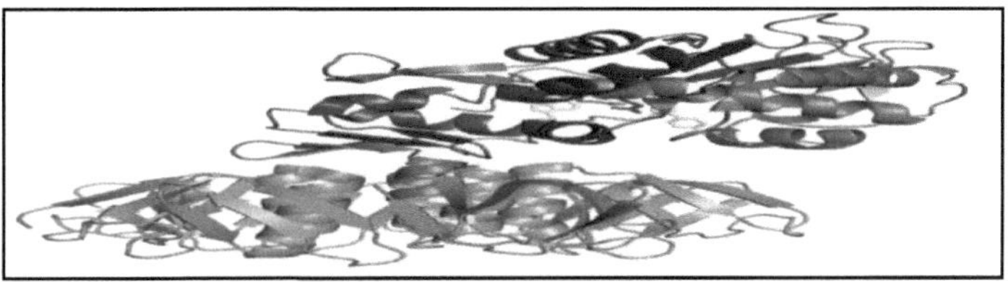

Figure 21: Crystal structure of Shiga-toxin 1.
The B pentamer is shown in orange and the A_2 in blue. The majority of A_1 is shown in green with the exception of the ribosome-interacting region, which is shown in purple. The active residue 167 is red and another active site in the side chains is pale blue.

c. Mechanism of action

In the first stage, the Stx1 toxin attaches to the cytoplasmic membrane of the target cell: the B subunits, assembled in a ring, bind to their receptor, the globotriosylceramide Gb3. Once the toxin has been internalized by the classic mechanism of endocytosis, it undergoes retrograde transport through the Golgi apparatus and then the endoplasmic reticulum. The A subunit is then split into two parts, A1 and A2. Part A1 possesses N-glycosidase activity which acts on the 28S rRNA of the 60S subunit of the ribosome, leading to excision of an adenine at position 4324. This depurination leads to ribosome inactivation and, consequently, the cessation of protein synthesis (Figure 22). The toxin (and in particular the B subunits) may induce cytokine production.

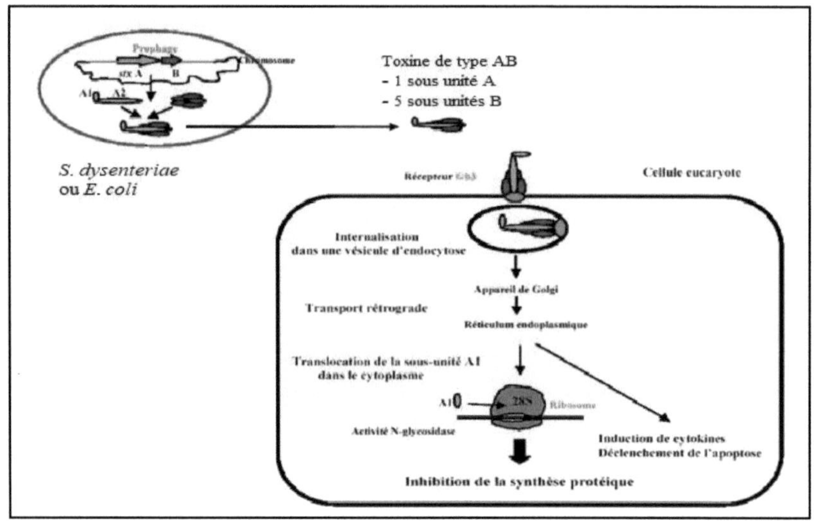

Figure 22: Mode of action of shiga-toxins.
Following binding of the toxin to the Gb3 receptor, internalization, retrograde transport and translocation, the N-glycosidase activity of the A1 subunit on 28S rRNA leads to total inhibition of protein synthesis.

d. Pathologies due to Stx

After crossing the intestinal epithelium, toxins are capable of systemic dissemination, and are carried to target organs by the bloodstream, either via red blood cells or polynuclear cells. They are responsible for thrombosis in local and systemic attacks, through alteration of the endothelial cells. Human vascular endothelial cells in the colon, renal parenchyma and central nervous system are particularly rich in Gb3 receptors, which explains the clinical manifestations observed (diarrhea, renal failure, neurological disorders).

A synergistic role for Stx and cytokines has been demonstrated: Stx induces the production of pro-inflammatory cytokines. This induces lesions in the intestinal barrier, leading to systemic dissemination of toxins and the development of hemolytic uremic syndrome.

III.2.4. Action on the nervous system

The nervous system is the body's regulatory and communication center, divided into two main parts: the central nervous system (CNS) and the peripheral nervous system (PNS). Nerve cells are the functional units of the nervous system, of which there are an estimated 10 billion.

III.2.4.1 Action by cleavage of SNARE proteins

SNAREs (Soluble N-ethylmaleinamid-sensitive factor Attachment protein Receptor) are proteins that control the fusion of synaptic vesicles with the plasma membrane of nerve endings. In effect, they act as tags, enabling mutual recognition and triggering fusion between vesicle and target membrane. They thus form the presynaptic fusion complex and participate in calcium-dependent exocytosis of neurotransmitters in the synaptic cleft.

These presynaptic SNAREs are made up of syntaxin 1, SNAP-25 in the plasma membrane and VAMP 2 (synaptobrevin) in the synaptic vesicle. Undeniably, the crystallographic structure of the SNARE complex suggests that its assembly would bring the vesicular and plasma membranes closer together, favoring mixing of the bilayers, and fusion.

SNAREs are also substrates for botulinum and tetanus neurotoxins, powerful inhibitors of exocytosis.

III.2.4.1.1. Example of botulinum toxin A

a. Definition

Botulinum neurotoxins (BoNTs), designated A, B, C, D, E, F and G according to their antigenic properties, are produced by various species of *Clostridium*. Bacteria belonging to this genus are strict, spore-forming anaerobic bacilli.

They are metalloproteases which cleave one of the three SNARE complex proteins playing a key role in the neuroexocytosis process, thereby inhibiting acetylcholine release at neuromuscular junctions.

b. Chemical structure

Botulinum toxin type A is naturally synthesized as a double-stranded protein of around 1300 amino acids (150 kDa). It is activated by proteolytic cleavage, which generates a light chain and a heavy chain that are joined by a disulfide bridge known as the belt (Figure 23). The light chain (50 kDa) is the catalytic domain of the toxin and carries a zinc-dependent endopeptidase function, while the heavy chain (100 kDa) consists of a translocation domain and a carboxy-terminal domain which serves as the receptor-binding domain. The protein's crystallographic structure showed that the three domains form distinct structural units.

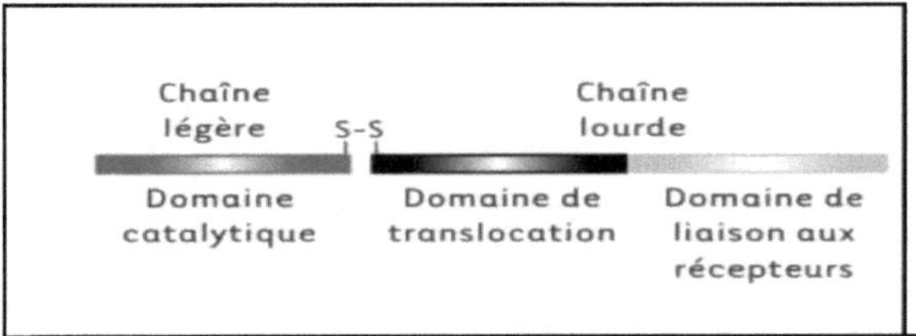

Figure 23: Schematic representation of BoNT and TeNT.

c. Mechanism of action

The molecular mode of action of BoNTs, which block the release of neurotransmitters from affected neurons, involves four steps:

❖ ***Connection to a receiver*** :

BoNTs pass through the digestive tract, crossing the intestinal barrier by a mechanism still poorly understood, then diffuse via the blood and/or lymph to cholinergic   motor neurons, where they bind to receptors on the neuron surface via the C-terminal binding domain of their heavy chain (Figure 24). These receptors are located on the amyelin regions of neurons, and are composed of gangliosides, synaptic vesicular proteins. BoNT-A receptors are synaptotagmin isoforms.

❖ ***Internalization*** :

BoNTs do not cross the cytoplasmic membrane directly. They are internalized by endocytosis into acidic intracellular compartments, and remain at the presynaptic terminals of motor neurons (Figure 24).

❖ ***Translocation into the cytosol*** :

Acidification of these endocytosis vesicles by the vesicular ATPase induces oligomerization of the neurotoxin heavy chains into tetramers, which insert themselves into the lipid bilayer, forming ion channels that allow the light chain to cross the vesicular membrane (Figure 24), localizing it in the cytosol after reduction of the disulfide bridge holding the two chains together, so that it can exert its activity.

❖ ***Enzymatic activity*** :

The substrates of BoNTs are SNARE proteins. Thus, botulinum toxin A is responsible for cleaving the SNAP-25 protein (Figure 24), leading to removal of the domain of SNAP-25 interacting with synaptotagmin which controls the Ca^{2+} ion flux required for neurotransmitter exocytosis.

As a result, the proteolytic action of BoNT-A causes synaptic vesicles not to fuse with the cytoplasmic membrane, preventing the release of acetylcholine into the synaptic cleft. Neuromuscular transmission is thus inhibited, leading to paralysis.

d. Botulism

Botulism is a fatal disease, also known as botulism poisoning. It is a serious paralytic disease caused by BoNT.

The toxin enters the body in one of the following ways: by colonization of the *C. botulinum* bacteria in the digestive tract in children or adults, by ingestion of the toxin from food, or by contamination of a wound by the bacteria, and leads to **paralysis** which usually begins with the facial muscles and then spreads to the limbs. In severe forms, it leads to **paralysis of the respiratory muscles** and **respiratory failure**.

This worldwide disease affects all ages, from infants to the elderly.

III.2.4.1.2. Example of tetanus toxin

a. Definition

Tetanus neurotoxin (TeNT) is an extremely potent toxin produced by *Clostridium tetani*, a strict anaerobic Gram-positive bacillus. TeNT is distinguished from other clostridial neurotoxins by its unique ability to target the CNS via retrograde axonal transport. It then inhibits the neurotransmission of inhibitory inter-neurons, causing spastic paralysis, including tetanus disease.

b. Chemical structure

Tetanus toxin is naturally synthesized as a 150 kDa polypeptide, which is post-translationally cleaved by a host or bacterial protease to generate a light chain and a heavy chain that remain linked by a disulfide bridge (Figure 23). The light chain (50 kDa) is composed of an N-terminal catalytic domain, a zinc-dependent protease, which specifically cleaves SNARE protein members. The heavy chain (100 kDa) consists of two 50 kDa domains, each with a distinct function, the N-terminal translocation domain responsible for light chain transport across the synaptic vesicle membrane into the cytoplasm. The C-terminal heavy chain domain is responsible for binding the toxin to peripheral neurons.

c. Mechanism of action

TeNT is a CNS neurotoxin. Since the toxin cannot cross the blood-brain barrier, it enters motor nerve endings without affecting acetylcholine release. It is internalized within intracellular vesicles with a neutral pH, and enters the CNS via retrograde axonal transport to the cell bodies of motor neurons, which synapse with inhibitory inter-neurons. This is where the toxin is released to exert its action on protein subunits of the SNARE system.

The VAMP protein (synaptobrevin) is the target of TeNT, and once cleaved by TeNT does not complex with membrane proteins (Figure 24). This makes it impossible for the neurotransmitter-containing synaptic vesicle to fuse with the plasma membrane. Exocytosis of glycine and γ-aminobutyric acid (GABA) by inhibitory interneurons is blocked. Motor neurons no longer receive inhibitory control. Muscle cells are therefore no longer properly regulated.

The result is muscular hypertonicity, first localized, then generalized, leading to the spastic paralysis characteristic of tetanus.

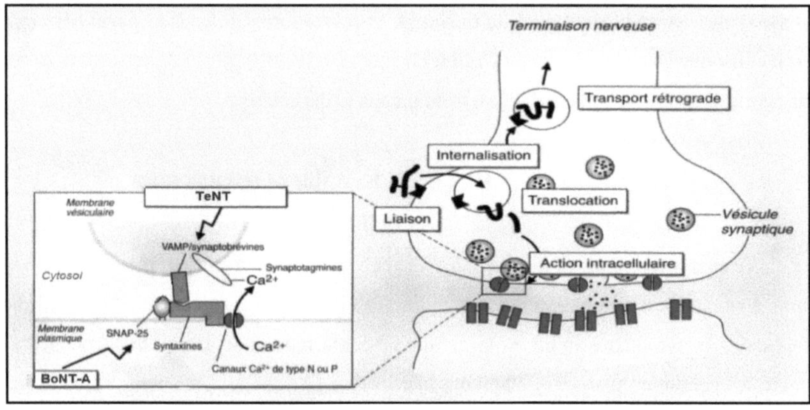

Figure 24: Intracellular action of BoNT and TeNT chains.

d. Tetanus

Tetanus means rigidity or tension in Greek, and is also a nervous disorder, characterized by spastic paralysis, i.e. contractions of the skeletal muscles (rigidity), and spasms (figure 25), often manifesting as dysphagia and spasms of the respiratory, laryngeal and abdominal muscles, causing respiratory insufficiency, and autonomic nervous system dysfunction.

Tetanus is now rare in most countries, but the <u>neonatal form</u> remains a major cause of mortality in certain regions, mainly in Africa. Prevention is based on vaccination and wound disinfection.

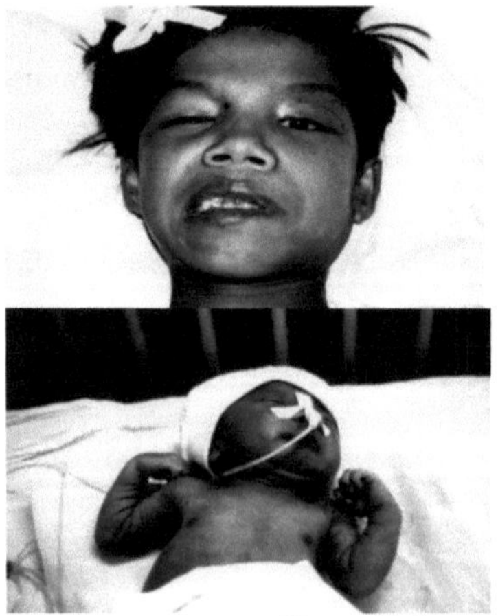

Figure 25: Some cases of tetanus.
Cephalic tetanus on the left and neonatal on the right.

III.3 Immunotoxins

This group includes toxins that interact with cell receptors linked to major histocompatibility complexes (MHC): these are immunotoxins or superantigens (SAg). In fact, they are single-stranded proteins, 20 to 30 kDa in size, capable of inducing polyclonal activation of T lymphocytes. This family of toxins includes the streptococcal toxic shock syndrome toxin from *Streptococcus pyogenes*, and the toxic shock syndrome toxin 1 (TSST-1) from *Staphylococcus aureus*.

III.3.1. Example of *Staphylococcus aureus*   TSST-1 toxin

a. Definition

Toxic Shock Syndrome Toxin 1 (TSST-1) belongs to a large family of exotoxins produced by *Staphylococcus aureus*. This super-antigenic, immuno-stimulatory toxin has the ability to interact with MHC class II (MHC-II) molecules and T-cell receptors.

b. Chemical structure

TSST-1 is a protein encoded by the *tstH* gene (15.2 Kb) present in the bacterial chromosome, which is transcribed and translated into a 234 amino acid precursor protein. This molecule is secreted after cleavage of a 40-amino acid signal peptide located on its N-terminal side. The mature protein is therefore a single-chain polypeptide of 22 kDa. The crystal structure of TSST-1 represents two domains, A and B (Figure 26).

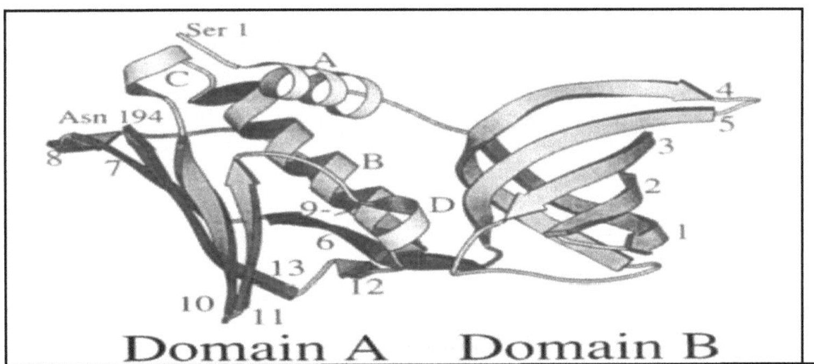

Figure 26: Ribbon drawing showing TSST-1 secondary structural elements in standard orientation.

The helices are labeled A-D and are shown as coils. β-sheets are labeled 1-13 and are represented by arrows.

c. Mechanism of action

SAg, including TSST-1, constitute a family of proteins that bind simultaneously to the α and/or β chains of MHC-II molecules and to the Vβ regions of T cell receptors (TCRs) (figure 27)   .

This binding leads to the activation of a large population of T lymphocytes, which release massive quantities of inflammatory cytokines, as well as the activation of macrophages to produce pro-inflammatory cytokines, including but not limited to tumor necrosis factors (TNF-α and TNF-β), interleukin 1β (IL-1β) and interleukin 2 (IL-2) and interferon gamma (INF-γ), resulting in toxic shock syndrome (TSS).

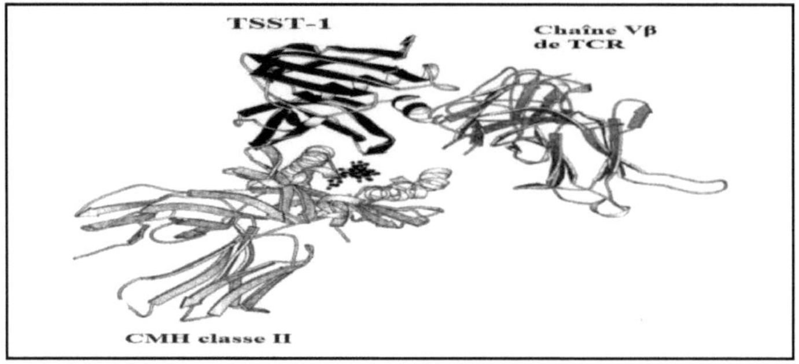

Figure 27: Structure of the MHC-II-TSST-1-TCR complex.

d. Toxic shock syndrome (TSS)

Staphylococcal toxic shock syndrome (TSS) is a potentially severe condition. It occurs suddenly after *S. aureus* enter the bloodstream and produce TSST-1. Various organs, including the liver, lungs and kidneys, are rapidly affected. This toxin triggers a systemic inflammatory cascade, favored by the absence or insufficient quantity of anti-TSST-1 antibodies.

Symptoms associated with TSS include flu-like symptoms such as fever, headache and muscle aches. These symptoms are rapidly progressive and severe. Other symptoms may include pain at the wound site, vomiting and diarrhea, signs of shock including low blood pressure and dizziness, shortness of breath and a sunburn-like rash.

IV. Mode of action of endotoxins

A growing number of studies indicate that endotoxins or lipopolysaccharides play a central role in a number of pathologies. Endotoxins are among the most toxic biological substances released by Gram-negative bacteria.

<h1 style="text-align:center">IV.1 Example of typhoid toxin from Salmonella enterica serovar Typhi</h1>

a. Definition

Typhoid toxin is an LPS molecule, a virulence factor of *Salmonella* Typhi, responsible for typhoid fever in humans.

b. Mechanism of action

After reaching the gastrointestinal tract, *Salmonella* Typhi rapidly invades the intestinal mucosa, even multiplying inside phagocytes and lamina propria, and then escaping. The bacteria also colonize the mesenteric lymph nodes, from where they diffuse into the lymph and then into the general circulation.

In addition, bacterial lysis in the lymph nodes releases toxic LPS or endotoxin O. Endotoxin O is carried in the bloodstream, causing local ulceration (perforation) of Peyer's patches (hemorrhagic intestinal lesions). Then, carried to the cerebral ventricles, it causes tuphos. This is a state of torpor, prostration, diarrhea and abdominal pain. In addition, endotoxin also reaches the thermoregulatory center (the hypothalamus), acting with phospholipase A2 to release arachidonic acid, which in turn releases prostaglandin via cyclooxygenase. This crosses the blood-brain barrier and acts on the sensory nerve, generating a plateau fever at 40°C, leading to typhoid and delirium .

d. Typhoid fever

Salmonella is frequently found in the intestinal tract of many animals. So, when <u>hygiene conditions are poor</u>, there's a risk of *Salmonella* contamination of food, and hence of humans. As a result, *Salmonella* bacilli enter the digestive tract and cross the intestinal wall. Bacteria penetrate the apical pole of cells during phagocytosis by macrophages.

Escape from the destructive mechanisms produced by the phagocyte will therefore be a key phenomenon in the virulence of this species, and is due either to Salmonella's resistance to enzymes, or to the fact that Salmonella effectively prevents phagolysosomal fusion. Bacteria are thus found within enterocyte vacuoles.

On the other hand, certain germs reach the lymph nodes, where they multiply and give rise to mesenteric adenolymphitis. In fact, the most obvious lesions are located in the Peyers' plaques of the terminal Ileum. From here, the bacilli enter the bloodstream via the lymph (Figure 28).

In addition, <u>bacterial lysis </u>within the <u>mesenteric lymph nodes </u>releases an <u>endotoxin</u>. It is the <u>release of this toxin </u>that is responsible for most of the clinical manifestations of these diseases.

It also causes complications through its action on the neuro-vegetative system, leading to typhoid fever.

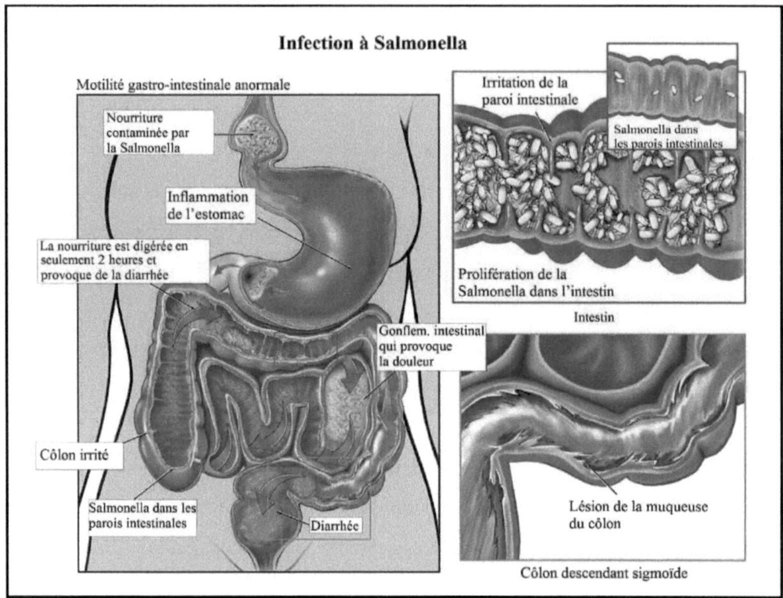

Figure 28: Representative diagram of *Salmonella* infection.

CHAPTER II: TOXINOTHERAPY

Toxins are virulence determinants that play an important role in microbial pathogenicity and/or escape from the host immune response. These molecules are therefore ideal targets for the development of new antimicrobial strategies. Bacterial toxins are therefore an active field of research, with implications not only for microbiology and cellular and molecular biology, but also for the medical field, specifically **toxin therapy**, in which bacterial toxins are used to treat certain conditions.

Indeed, the potential applications of toxin research go beyond simply combating microbial pathogens, and include use as new anti-cancer drugs and neurobiological tools.

A. Anatoxins

I. Definition

In 1924, Ramon described that toxins such as TeNT and DT could be inactivated by heat and formaldehyde treatment, giving immunogenic but non-toxic forms (anatoxins).

This discovery paved the way for the production of the first vaccines by chemical treatment. A toxoid contributes to the secretion of immune cells capable of detecting the toxin. So, by injecting a patient with a toxoid, his or her immune system will be all the more reactive once it comes into contact with the toxin.

II. Preparation of toxoids

Formalin-induced toxin detoxification, following Ramon's method, appears to retain its immunological properties essentially intact. Thus, the loss of toxicity could derive from the preferential modification of the region(s) involved in the expression of this activity. In other words, it may involve a modification of one or more amino acid residues located in the toxic site.

The production of a bacterial toxoid involves several stages. It begins with the growth of specific strains. These bacteria are first grown on solid media, then placed in liquid culture for several days to a week, to adapt to the medium and develop. Temperature and culture conditions are tightly controlled to prevent contamination.

Once the bacteria have developed, the culture is harvested and the cells removed by centrifugation and/or filtration, allowing the secreted toxin to be recovered.

The toxin is treated with a chemical agent such as formaldehyde, thus eliminating its toxicity, while retaining the protein structure needed to trigger a protective immune response. The resulting molecule is called <u>anatoxin</u>.

The toxoid is purified by a variety of processes, including precipitation (addition of a salt causing selective precipitation of toxoid), electrophoresis and/or chromatography. The toxoid is then tested for purity, safety (absence of toxicity) and potency before formulation into the final vaccine.

III. Example of the tetanus vaccine

Tetanus toxoid (TT) is obtained by inactivating the toxin with formalin. AT requires the use of adjuvants, the most common being aluminum salts (aluminum hydroxide or aluminum phosphate) on which it is adsorbed, to enhance its antigenicity.

Immunity to tetanus is a humoral immune response. Indeed, tetanus toxoid induces the formation of specific antibodies belonging to the IgG class. These antibodies can easily cross the placental barrier and diffuse throughout the bloodstream and extravascular zones.

In addition, studies of neonatal tetanus mortality in children born to both vaccinated and unvaccinated mothers provide information for assessing the efficacy of the tetanus vaccine. In most of these investigations, this efficacy is between 80 and 100%.

In addition, vaccination of mothers can protect the fetus against neonatal tetanus, thanks to the passage of antitoxins (IgG) across the placental barrier.

Figure 29 shows antitoxin titres in adults after primary and booster vaccination with tetanus toxoid. The level and duration of immunity increase with the number of toxoid injections.

Two to four weeks after the second injection, the average level of anti-tetanus antibodies exceeds the minimum protective level of 0.01 IU/ml.

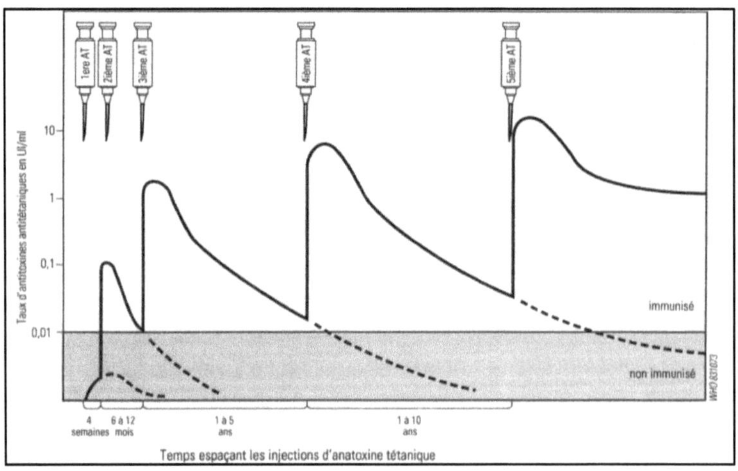

Figure 29: Antibody response to tetanus toxoid

B. Use of botulinum toxin

Botulinum toxin was the first toxic protein derived from bacterial cultures to be used therapeutically. In fact, the toxin's neuronal specificity has made it a therapeutic tool used in a wide range of indications in physical and rehabilitation medicine, neurology, ophthalmology, urology and pain management.

There are currently four botulinum toxin specialties. These are three type A botulinum toxins: BOTOX®, DYSPORT®, XEOMIN® and one type B botulinum toxin: NEUROBLOC®. The best standardized is type A.

I. Aesthetic treatment with BOTOX

I.1 Definition and principle

BOTOX is a trade name for "botulinum toxin", which comes in the form of a purified protein. It is widely known for its use in cosmetics to reduce wrinkles. It works by blocking neuromuscular transmission, resulting in a temporary relaxation (reduction) of tone in the injected muscle. In this way, it reduces wrinkles by inducing a transient, reversible paralysis of the treated muscle.

In fact, BOTOX specifically treats so-called **dynamic** wrinkles, also known as expression or contraction lines. These are located in the expression zones (forehead, eye contour, mouth contour) where repeated muscular contraction causes "waves" in the skin. In other words, the repetition of facial movements (frowning, wrinkling, smiling...) gradually creates marks on the face. So, over time, wrinkles form and remain present even when muscles are relaxed. This class includes :

o Wrinkles around the eyes, known as "crow's feet", express joy.

o Glabellar lines, or "frown lines", are the vertical wrinkles between the eyebrows. They express anger.

o Transverse forehead wrinkles express astonishment.

o Perioral wrinkles or smile lines.

In addition, BOTOX, which specifically treats this type of wrinkle, gives the treated patient a totally relaxed, more rested and therefore younger-looking face.

Unquestionably, BOTOX treatment cleaves the SNAP25 protein, blocking acetylcholine release and preventing muscle contraction.

This partial denervation of muscles results in a localized reduction in muscle activity. The effect on the skin is to smooth out contraction wrinkles.

This blockage is irreversible, but the nerve endings will "grow back" and re-establish other connections with the muscle, which will then re-inner itself.

Full recovery of motor function/nerve conduction normally occurs within 3 to 4 months, hence the need for repeated injections.

I.2. Historical background

Botulinum toxin is a protein complex produced from *Clostridium botulinum*, a gram-negative anaerobic bacterium discovered in Germany in the 19ème century by Justinus Kerner as the cause of food poisoning due to the consumption of black pudding. Thus, "botunlinum" comes from the Latin "botulinus", meaning "blood sausage".

It was then studied in the 1970s by Alan Scott, an American ophthalmologist, who, with the help of a biochemist, developed Botox as a treatment for ocular spasms and strabismus, for which he obtained FDA approval in 1989.

Its use for eye disorders led Dr. Carruthers, a Canadian ophthalmologist, to note that one of its side effects was the reduction of vertical wrinkles between the eyebrows. In 1990, she introduced the first cosmetic use of botulinum toxin to treat facial wrinkles.

In 2002, Botox obtained FDA approval for use in glabellar wrinkles, and in 2013 for the treatment of crow's feet.

I.3 BOTOX preparation

Since its initial use in ophthalmology, BoNT has been produced from an American strain of *C. botulinum* known as Hall, selected for its high concentration of BoNT production (over 10^6 IU/ml at 37°C) in 24 to 36 hours.

Bacterial culture in a suitable medium is carried out under anaerobic conditions. The BoNT is then purified and diluted in a solution of lactose and human serum albumin, enabling stabilization prior to galenic packaging in the form of a lyophilisate to be stored at a temperature below -5°C. Dilution in physiological saline conditions local diffusion. After reconstitution, the solution should be used within 4 hours of opening, provided it is stored between +2 and +8°C.

I.4. Treatment of facial wrinkles

The treatment of wrinkles with Botox is simple to perform, and has proven to be both safe and effective in the short term. It involves local injections directly into the desired muscle, using small syringes or fine needles. The injection is painless.

The most common indication is the glabella, between the 2 eyebrows. BOTOX injections are also used for crow's feet and forehead wrinkles, and to raise the eyebrow.

Botulinum toxin injection induces muscle relaxation between 2ème and 10ème days, with optimal "smoothing" results after 3 weeks. The smoothing of wrinkles lasts between 4 and 6 months, and the operation should be repeated after this period or as soon as new wrinkles appear (Figure 30).

Although the injection itself only takes a few minutes, it does require a few precautions:
- ➢ The skin must be completely clean and makeup-free on the day of the injection;
- ➢ The practitioner traces out the injection points on the face;
- ➢ Careful dosage of botulinum toxin;
- ➢ Do not take anticoagulants in the days preceding the injection.

If all these conditions are met, a plastic surgeon can proceed with the Botox injection. Although not very painful, a few precautions should be taken after the injection to optimize the result:
- ➢ Apply an anaesthetic ointment to reduce pain;
- ➢ Do not touch or massage treated areas for 24 hours after injection;
- ➢ Do not exercise;
- ➢ Do not take aspirin or anti-inflammatory medication for 3-4 days to avoid spreading the product to unwanted areas.

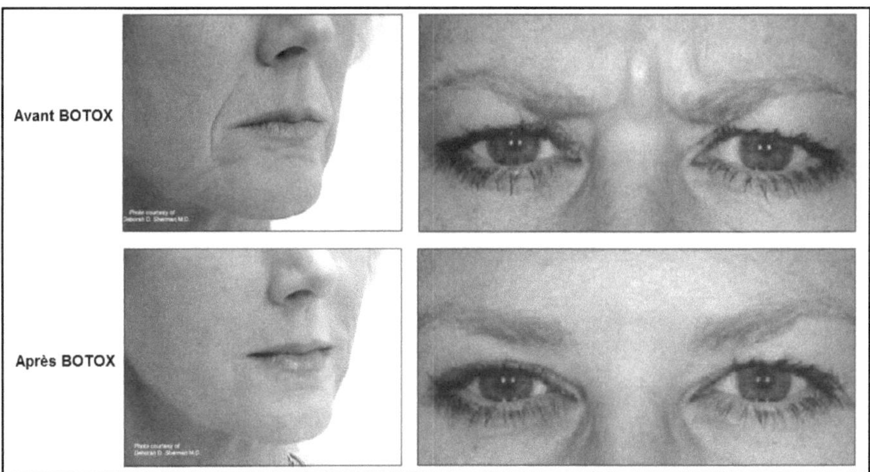

Figure 30: Images illustrating the treatment of facial wrinkles using BOTOX

C. Immunotoxins: a new tool for treating cancer

In recent years, several immunotoxins targeting tumors such as breast, lung, stomach, bladder and central nervous system cancers have been developed. Recently, they have been clinically tested on haematological malignancies and solid tumours, demonstrating potent clinical efficacy in patients with malignancies refractory to surgery, radiotherapy and chemotherapy, the traditional methods of cancer treatment.

Efforts are underway to eliminate obstacles to clinical efficacy, including immunogenicity and toxicity to normal tissues.

I. Preparation of DT immunotoxin

Toxins can be manipulated to kill cancer cells, simply by replacing the toxin's binding domain with an antibody, growth factor or cytokine (figure 31), which would selectively bind to the target cell.

Alternatively, the binding domain may be present but mutated to prevent its binding to normal cells.

Ligand and toxin can be linked chemically, as in chemical conjugates, or genetically, as in recombinant fusion toxins.

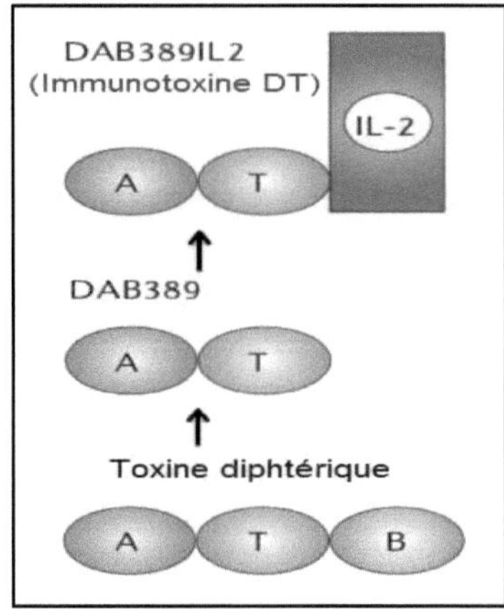

Figure 31: Strategy used to manufacture DT immunotoxin
(B): receptor-binding domain; (T): cytosolic translocation domain; (A): catalytic domain.

II. DT immunotoxin cancer therapy

Immunotoxins are designed to bind to specific antigens on the surface of cancer cells and enter the cells by endocytosis (Figure 32). Undoubtedly, endocytosis transports the toxin into an acidic compartment of the endosome, causing a structural change in the DT immunotoxin, enabling its translocation across the endocytic membrane.

Consecutively, the catalytic A domain ADP-ribosylates elongation factor 2 (FE2). FE2 thus modified by ADP-ribosylation can no longer produce new proteins, and the cell dies by apoptosis.

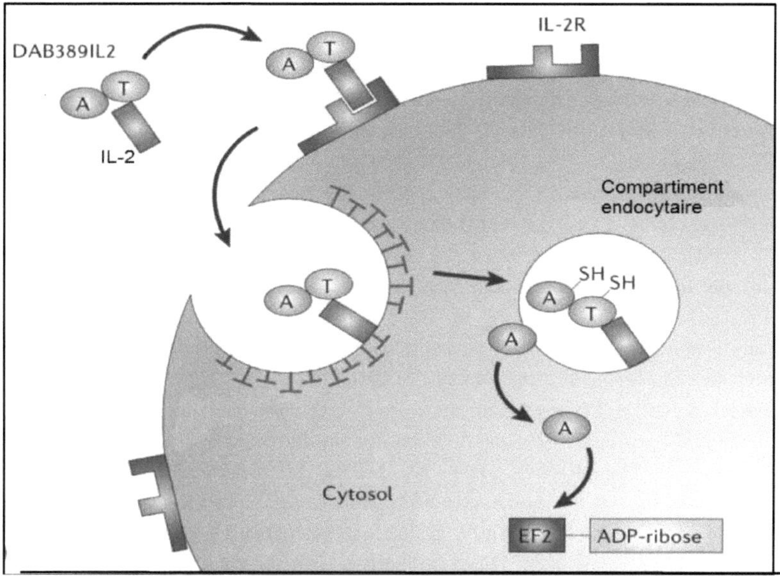

Figure 32: Action of DT immunotoxin on cancer cells.

It should be noted that the affinity of the immunotoxin for the antigen determines the binding time during which it remains associated with the antigen on the cell surface. Consequently, the higher the affinity, the more active the immunotoxin is in destroying target cells.

BIBLIOGRAPHICAL REFERENCES

- Baker M.D., Acharya K.R. 2004: Suprantigens: Structure-function relationships. *Int. J. Med. Microbiol.* 293: 529-537.
- Barth H., Fischer S., Möglich A., Förtsch C. 2015 : Clostridial C3 toxins target monocytes/macrophages and modulate their functions, Frontiers in immunology. 6: 1-6.
- Benz R. 2016 : Channel formation by RTX-toxins of pathogenic bacteria: Basis of their biological activity, Biochimica et Biophysica Acta, 1858(3): 526-537
- ChenalA., Nizard P. and Gillet D., 2002 : Structure and Function of Diphtheria Toxin: from Phathology to Engineering. Journal of Toxicology-Toxin Reviews. 21(3): 321-359
- Chong A., Lee S., Yang Y.A., Song J. 2017: The Role of Typhoid Toxin in *Salmonella* Typhi Virulence. Yale Journal of Biology and Medicine.90: 283-290.
- Deloye F., Schiavo G., Doussau F., Rossetto O., Montecucco C. Poulain B. 1996: Modes of molecular action of botulinum and tetanus neurotoxins. médecine/sciences. 12 : 175-182.
- Galmiche A., Bouquet P., 2001 : Toxines bactériennes : facteurs de virulence et outils de biologie cellulaire. médecine/sciences.17 : 691-700.
- Hassel B., 2013: Tetanus: Pathophysiology, Treatment, and the Possibility of Using Botulinum Toxin against Tetanus-Induced Rigidity and Spasms. *Toxins.* 5: 73-83.
- Hochedez P., Hope-Rapp E., Olive C., Nicolas M., Beaucaire G., Cabié A., 2010: Bacteremia Caused by *Aeromonas hydrophila* Complex in the Caribbean Islands of Martinique and Guadeloupe. Am. J. Trop. Med. Hyg. 83(5) :1123-1127
- Hotze E M., Tweten R K. 2012: Membrane assembly of the cholesterol-dependent cytolysin pore complex, Biochimica et Biophysica Acta. 1818: 1028-1038.
- Mateyak M.K., Kinzy T.G. 2013: ADP-ribosylation of Translation Elongation Factor 2 by Diphtheria Toxin in Yeast Inhibits Translation and Cell Separation. Journal of Biological Chemistry. 288(34): 24647-24655 p.
- Ménager C., Kaibuchi K. 2003 : Les protéines Rho : leur rôle dans les neurones, Médecine sciences. 19 (3): 358-363.
- Pastan I., Hassan R., FitzGerald D.J., Kreitman R.J., 2006: Immunotoxin therapy of cancer. Nature Reviews/CANCER. 6: 559-564.
- SeagarM., Quetglas S., Iborra C., Lévêque C. 2001 : The SNARE complex at the heart of membrane fusion. médecine/sciences. 17 : 669-674.
- Spaan A. N., Schiepers A., De Haas C. J. C., Van Hooijdonk D. D. J. J., Badiou C., Contamin H., Vandenesch F., Lina G., Gerard N. P., Van Kessel K. P. M., Henry T., Van Strijp J. A. G., 2015: Differential Interaction of the Staphylococcal Toxins Paton-Valentine Leukocidin and γ-Hemolysin CB with Human C5a Receptors, The Journal of Immunology. 195: 1034-1043.
- Van Bost S., Mainil J., 2003: Virulence factors and specific properties of invasive *Escherichia coli* strains. *Ann. Méd.* 147: 327-342.

LEXICON

Apathy: is the inability to react.

Apnea: is a momentary suspension of breathing.

Septic arthritis: Arthritis is defined as acute or chronic inflammatory damage to the synovial mucosa of one or more joints. Septic arthritis occurs when an infectious agent, whether bacterial or fungal, multiplies in the joint synovial fluid.

Bacteremia: Transient passage of germs into the bloodstream.

Cholangitis: is a disease affecting the bile ducts and liver; it is an inflammation of the bile ducts that transport bile from the liver to the gallbladder and then to the intestine.

Acute cholecystitis: inflammation of the gallbladder. It generally occurs when the gallbladder duct is blocked by gallstones. Gallstones are more common in women, the elderly and the overweight.

Chorio-amniotitis: is an acute inflammation of the placental membranes and chorion.

Actin cytoskeleton: The cytoskeleton is a filamentous network inside a cell, giving it its mechanical properties.

Respiratory distress: is a type of respiratory (pulmonary) failure.

Granuloma: is a chronic proliferative inflammatory lesion consisting of mononuclear cells (mononuclear lymphocytes and macrophages) and, inconsistently and in variable numbers, epithelioid cells and multinucleated giant cells.

Eye infections: an infection of the eyes.

Meningitis: is an inflammation of the "meninges", the thin membranes enveloping the brain and spinal cord.

Myositis: is an inflammatory disease of muscle tissue. It is one of the musculoskeletal disorders.

Nephropathy: is a kidney disease caused by a functional disorder or organic lesions of the kidney.

Necrosis: is a process triggered when a cell is subjected to structural or chemical aggression from which it cannot recover, leading to its death.

Operon: An operon is a functional DNA unit grouping genes that operate under the signal of the same promoter.

Osteomyelitis: is a bone infection caused by a germ that reaches the bone via the hematogenous route. It occurs mainly in the metaphyses of long bones.

Peritonitis: inflammation of the peritoneum, the membrane that lines the inside of the abdominal cavity.

Phalloidin: a toxin derived from the *Amanita phalloides* fungus. By binding to actin filaments, the toxin inhibits their depolymerization, causing their accumulation and thus cell dysfunction. The toxic effect is essentially due to renal and hepatic damage.

Pneumonia: is an infection of the lungs, most often caused by a virus or bacteria.

Nosocomial pneumonia: develops in hospitalized people living in care centers or in contact with medical environments.

Prostration: this refers specifically to a state of extreme weakness and fatigue, manifested by the collapse of the patient's muscular functions and immobility.

Rhomboencephalitis: is a neurological condition corresponding to inflammation of the brain stem.

Septicemia: Or sepsis, **is** a generalized infection of the body of bacterial origin. It is manifested by repeated discharges of pathogenic germs into the bloodstream from an infectious site.

Sterane: A quadricyclic alkane consisting of three 6-atom rings (A, B and C) and a 5-atom ring (D). This structure has 6 asymmetrical carbons.

Synovia: stringy liquid that lubricates moving joints.

Torpor: General numbness, both physical and mental, which leads to semi-consciousness, drowsiness and drowsiness.

Tuphos: The state of stupor and extreme despondency that characterizes typhoid fever and other diseases such as typhus.

MIX
Papier aus verantwortungsvollen Quellen
Paper from responsible sources
FSC® C105338
FSC
www.fsc.org

Printed by Books on Demand GmbH, Norderstedt / Germany